Hutton's Arse

Other earth science titles from Dunedin include:

The Western Highlands of Scotland (2019)
Classic Geology in Europe
Con Gillen
ISBN: 9781780460406

Farewell, King Coal: from industrial triumph to climatic disaster (2018)
Anthony Seaton
ISBN: 9781780460772

Volcanoes and the Making of Scotland (2015)
Second edition
Brian Upton
ISBN: 9781780460567

Scottish Fossils (2013)
Nigel Trewin
ISBN: 9781780460192

For further details of these and other Dunedin
Earth and Environmental Sciences titles see
www.dunedinacademicpress.co.uk

Hutton's Arse

3 billion years of extraordinary geology in Scotland's Northern Highlands

Second edition

Malcolm Rider
and
Peter Harrison

EDINBURGH / LONDON

This Second Edition published 2019 by
Dunedin Academic Press Ltd
Hudson House
8 Albany Street
Edinburgh EH1 3QB
Scotland

www.dunedinacademicpress.co.uk

First edition published by Rider-French 2005
Reprinted 2007, 2011, 2012 and 2014

ISBNs
9781780460932 (Paperback)
9781780466088 (ePub)
9781780466095 (Kindle)

British Library Cataloguing in Publication data
A catalogue record for this book is available from the British Library

Designed by Melissa Alaverdy
Printed in Poland by Hussar Boooks

Contents

About the authors —————————————————————————— vi

Introduction —————————————————————————————— vii

1. A RED EARTH ————————————————————————————— 1
    The enigma of the ancient Torridonian

2. DEEP SCAR: THE MOINE THRUST ——————————————— 39
    The story of a bitter scientific controversy

3. THE FISH GRAVES OF ACHANARRAS ————————————— 73
    Devonian fish are human ancestors

4. VOLCANO ——————————————————————————————— 105
    The Tertiary volcanic province and the Atlantic opening

5. THE COMING ICE AGE ————————————————————— 139
    Past and future climates in the Highlands

6. LEFT-OVERS ———————————————————————————— 173
    The Lewisian Gneiss and the creation of continents

7. HUTTON'S ARSE (1795) —————————————————— 199
    Just another planet?

8. THE FUTURE ————————————————————————————— 217

    Index ———————————————————————————————— 221

About the authors

For over 30 years Malcolm Rider has been involved in the hydrocarbon industry, working first for a major company and then running his own consultancy business. He is still very much active in the industry, travelling the world presenting specialist courses and advising on oil exploration techniques. Peter Harrison has lived for 40 years in the north west Highlands working as a teacher and geoscience educator. He wishes to share the inspiration of this thought-provoking landscape; Peter leads geotours for the North West Highlands UNESCO Global Geopark.

The cover photograph is of the statues of Ben N Peach and John Horne at Knockan Crag National Nature Reserve. They played the foremost part in unravelling the geological structures of the North West Highlands, 1883–1897. The North West Highlands are a UNESCO Global Geopark.

Introduction

This book is about rocks – their science and its history – but also the people, the stories, the places and the secrets they hold. It is equally about the magical northwest Highlands of Scotland, still startlingly remote but long lived-in. All of them – place, land, rocks, history, people and science – are inseparable, none adequate without the other, all intricately part of each other. Beneath its surface of science this book also blunders into the many colours of emotion, ambition, jealousy and the arrogance of clever people. We have much to learn from this remote and sparsely populated area. It takes a while to absorb its offering, so why not come and see, listen and feel what it has to tell you, and let this book act as an introduction to a journey that has so much to offer.

The northwest Highlands of Scotland more than deserve to be the inspiration for a book on geology. The northern climate has created a mystical scenery of bare and barely covered rocks and crags and has long attracted geological researchers. The science in this book is serious and frequently at the forefront of present research, but it cannot and should not be separated from the place. Stand on any one of the peaks along the western coastal range and look into the wind, towards the Atlantic: what you see is the result of three billion years of geology. It is such a memorable panorama. Across your view a lacework plateau of land and water slips westwards imperceptibly into a sea studded with ragged promontories and small islands. It is as if the land is being slowly dissolved at its edges. On to this plateau Suilven, Cul Mor, Cul Beag, Stac Pollaidh and Quinag are placed like the last few pieces at the end of a chess game: abrupt, isolated, strangely distributed rock monuments, sloping and firm at the base but cliffed and sculpted above as the individual chess pieces they are: Suilven the King, Stac Pollaidh the Queen. For the others, come north and see for yourself.

Popular science books today often have an attitude problem. Some are written as entertainment and excessively dumbed-down. Some take the posture of teacher to pupil, superior to inferior, which is especially the case with books concerning environmental matters that tend towards sermonizing, threats and moralizing. 'You must stop driving your car and eating meat or you will destroy the rain forests and cause global warming.'

It cannot be too extensively known that nature is vast and knowledge limited, and that no individual, however humble in place or acquirement, need despair of adding to the general fund.

This is how, in 1841, Highland geologist Hugh Miller viewed science; a general playground for all who were interested in it. He was quite right. Not only did Hugh Miller love looking at rocks and fossils but he wanted to share his pleasures with others – sentiments that would sit well in science today. This is not science that is only read in books or stared at in a video. This is science that can be touched and experienced, science that lets you do your own exploring, lets you take your own decisions and, especially, make your own discoveries: science as it should be. And this is the excuse for mentioning Hutton's arse (which he himself also did, and on many occasions).

Although he was a consummate theorist, Hutton was most definitely a field geologist and his backside went through discomfort because of it (the only way for him to explore the rocks was on horseback). His ideas came from what he saw. What Hutton saw, though, was not what others saw. He was the only one at the time to believe in the evidence of the rocks. Where some theorized water, Hutton observed fire. Where others wanted biblical convulsions and no time at all, he observed slow processes applied over a very long time. For his inspiration Hutton used the rocks, which he called 'God's little diaries'. So, although this book is about the science of modern geology, it is at the same time a guide to the rocks, the evidence and some of the history behind that science.

Discoveries and advances in science today are published in scientific journals. There are many; they are highly specialized and are read only by experts and academics. The articles are peer reviewed and extremely serious. There are strict rules as to how such scientific papers should be written. The first person is never used, emotional writing is not tolerated and even an exclamation mark will be removed! If a scientist is absolutely the only person to adhere to a particular idea, he or she is still not allowed to write 'I believe'; it has to be the less exciting (and less truthful) 'it is believed that'. The science may be good but the writing has become turgid, boring and difficult to read. To break out of these restrictions professionals are turning to writing books and publishing more personal and readable accounts of their work. We are returning to the days when scientists paid for their own works to be printed, just as James Hutton did in 1795 for his *Theory of the Earth with proofs and illustrations.*

It is impossible to live in Scotland today and not be influenced by Edinburgh, the country's beautiful capital on the crags. Between 1750 and 1800 Edinburgh became the Athens of the North, the city of the 'Enlightenment', a ferment of intellectual activity and the time of James Hutton and his many illustrious friends. In 1997, 290 years after Westminster had taken it away, Edinburgh had a Parliament once more and even the Stone of Destiny returned to the north. The city is invigorated again, not thanks to its politicians but through its writers, artists, musicians, media people and even scientists. Edinburgh City in the early twenty-first century is a great and lively place to be and has been an inspiration. But there can be no doubt that the overwhelming influence on the formation of this book has come from the Highlands themselves.

The far northwest of Scotland has magic in its scenery, astounding geology, and is almost empty of people. For the moment it is a beautiful place in which to breathe, feel and think, and if we can share this, explain some geology and communicate some feelings, that will do.

To recognize the special quality of the area, the North West Highlands UNESCO Global Geopark was first established in 2004. Much of what lies within these covers is inspired by the story to be told from the rocks within and just outside the geopark boundary. The geopark aims to celebrate scientific understanding and progress whilst at the same time supporting the sustainable development of the area; for there is little point in a land of such wonders if there is no resident population to interpret the landscape and culture for the increasing numbers of visitors. The website www.nwhgeopark.com could be the starting point to explore the very land that has so much to tell us.

To understand how the Earth works is a challenge to us all. It is a journey, and a very enjoyable one at that. It should not just be viewed as a pleasure for its own sake, however, as the decisions we all need to take about our future tenure of the Earth rely on such an understanding. The people of the United Kingdom, and Scotland in particular, have a wealth of varied landscapes to help foster that understanding, and the northwest Highlands overflow with potential to test our powers of interpretation. Accept the challenge and find out more.

An excellent companion within the geopark area to explore the geology of the northwest Highlands is *A Geological Excursion Guide to the North-West Highlands of Scotland* edited by Kathryn M Goodenough and Maarten Krabbendam, published in 2011 by

Edinburgh Geological Society in association with National Museum of Scotland Enterprises, ISBN 9781905267538. This explores the geology of the geopark area including maps, photographs and descriptions of identified localities. For a wider geographical area with a comprehensive account a recently published title is recommended: *The Western Highlands of Scotland* in the series Classic Geology in Europe 9, 2019, Dunedin Academic Press, Edinburgh, ISBN 1903544 173 by Dr Con Gillen. This includes descriptions of localities to see the geology and an excellent further reading list.

A series of booklets has been written by Alan McKirdy called Landscapes in Stone. *The Northern Highlands*, ISBN 9781780276083, is the most recent to join nine others.

Chapter 1 A RED EARTH

The enigma of the ancient Torridonian

rom the seven houses that make up the small west coast Highland village of Stoer, you can turn your back on the sea and look over the land eastwards, and in the low light of a late spring evening, watch the sun and shadows play on that surprising peak of Suilven. The hill rises with suddenly steep walls, like a sand castle being destroyed in the tide, so that as the evening light slowly changes, unlikely streaks are lit up and ominous, illogical shadows left behind. It is a light show, and for stage effects Prospero could not ask for better. It is not a usual hill, Suilven, neither in its shape nor in its rock. For rock it surely is. But to talk of Suilven is to talk of the sister hills of Canisp, Cul Mor, Cul Beag and Quinag, all ancient Gaelic titles. So remarkable is the presence of these hills that something profound is quite expected of them. Just as a great cathedral demands silence, these hills provoke an expectation. Nor is it disappointed; for here it is, amongst these hills, that geological time itself is being dramatically cheated.

It was perhaps on such a spring evening early in the 1800s that Dr John Macculloch (1773–1835) first sailed into the Bay of Stoer and was so impressed by this sight from the sea. English-born but trained in Edinburgh, a medical doctor and a chemist but in his heart a geologist, Macculloch was one of the first to apply science to the landscape of the Scottish far northwest, looking for limestone on his early visits and then later working for the government Trigonometrical Survey. A dramatic geological sketch that he made in 1819, from the sea looking inland, of the hills between Quinag and Cul Mor, including of course Suilven, clearly shows his understanding of the rocks. To him must go the credit for stimulating other, mainly absentee English geologists, to visit the area. The new path up Suilven from Canisp Lodge would have amazed him.

What impressed Macculloch was the fact that horizontally layered sedimentary rocks existed in this area at all. He knew from his own and others' work that the local rocks were old and crystalline and often fractured and distorted. And yet here, clear to see, were almost undisturbed layers. We know now, of course, that what he observed so accurately were the exceedingly ancient, even though undisturbed, Pre-Cambrian, red Torridonian sediments, products of ancient deserts, long gone rivers and ephemeral lakes. But these cover even more ancient rock, Archaean Lewisian Gneiss (pronounced 'nice') that are crystalline and distorted, characteristics inherited from their past deep within the Earth, perhaps as much as 40 to 50km down. The junction between the two, the sedimentary and the crystalline rocks, indicates a huge passing of time, a geological unconformity, as there must always be when once molten and subsequently contorted rocks are quietly smothered by waterlaid sediments. Even in the early 1800s this was known (thanks to James Hutton, Chapter 6) and Macculloch was well aware of it, as his sketch shows. But he was only aware of the superficial facts; he did not have the modern tools of geology that can tease much more information out of the rocks than he ever could have dreamed. This chapter is going to look at the ancient clues in the Torridonian. Read the runes and they tell us what the ancient,

Figure 1.2
John Macculloch's 1819 sketch from near Stoer looking east. From left to right (north to south) Quinag, Suilven, Cul Beag and Cul Mor. The Torridonian–Lewisian unconformity is very clear (after Macculloch, 1819).

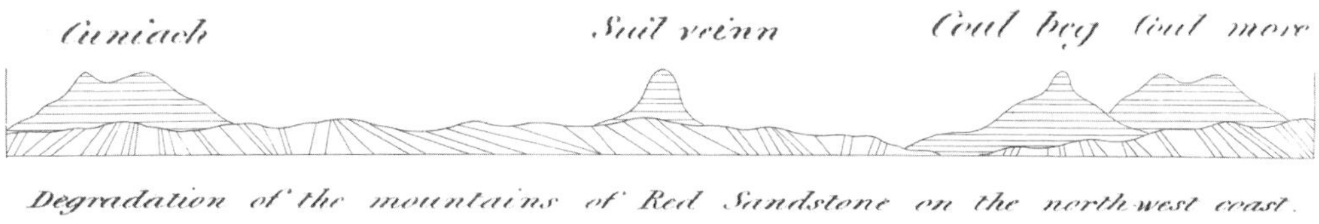

Cuniach　　　　　*Suil veinn*　　　*Coul beg　Coul more*

Degradation of the mountains of Red Sandstone on the north west coast.

Pre-Cambrian Earth would have been like: the rain, the heat, the wind, the seas, the life (primitive, but life) and the catastrophes.

Before beginning to look at the rocks themselves, we must introduce two modern specialist branches of geology. The first is the study of the sediments themselves, that is, sedimentology, one of the subjects of this chapter. The name does not matter. The specialists who use sedimentology carefully examine rivers, beaches, deltas and all the places where sediments collect today, in order to be able to recognize them fossilized in rocks. For example, the ripples that we see in the sand by the seashore are found in solid slabs of rock identical to those we leave a footprint in. Modern science yields much, much more than just simple comparisons like this, of course.

The second speciality is stratigraphy, which has for its main task putting sediments in order of age wherever they are in the world. For the last 150 years, since geology became a science, it has been known that the age of a sediment can be recognized by its fossils, although in the last 50 years absolute ages have also been added. For example, Jurassic sediments are recognized by, amongst others, the fossil dinosaurs they contain; but we also know now that these sediments are between 145 million and 200 million years old. The chart used to record these ages, and memorized by all geologists, is called the 'stratigraphic column', and represents a compendium of all possible rock sections from around the world, clipped together into a universal set. The stratigraphic column, which is looked at again in Chapter 3, depends on the presence of fossils and works for sediments younger than 'only' 540 million years – that is, from the beginning of the Cambrian. Before this, when there was not enough or the wrong kind of life to be fossilized, we are dependent on absolute age dates that are not always available. Most familiar sediments are younger than Cambrian and old, unfossiliferous, Pre-Cambrian sedimentary rocks are rare. The reason for mentioning them here is that the red Torridonian sediments at Stoer, those horizontally layered strata that impressed John Macculloch, are just such very old Pre-Cambrian sediments; they are between 1200 million years and 800 million years old!

Reading 500 million years or 1000 million years is rather like being told someone has a fortune of 500 million or 1000 million – what is the difference? Millions are dreadfully rich and millions are unbelievably old. How can we judge these geological millions? Until very recently, geological time effectively stopped, or I should say started, at 540 million years ago (abbreviated to 540Ma, mega anna), at the base of the Cambrian when hard-bodied fossils first became common. A relatively recent book has a title that suggests it is about the geological timescale, though in reality it is only about the Phanerozoic, that is the time between the Cambrian and today. For its 0.54 billion* (0.54Ga, giga anna, i.e. 540Ma) years of time, the Phanerozoic gets nearly 120 pages in this book; for its 4.0 billion years (4,000,000,000 or 4.0Ga), that is 8 times as long, the Pre-Cambrian only gets a page and a half. This error is slowly being realized, but still today in most geological textbooks, below the Cambrian is an uninteresting, unfossiliferous gap, the Pre-Cambrian. The turning point in attitudes was certainly the dating of the Moon's surface at 4.45Ga and the realization that the Earth is even older. Since then, this enormous timespan before the Cambrian has received more and more attention and there has even been a change in the spelling. The Pre-Cambrian is now written Precambrian, although as Romeo says '*what's in a name?*'

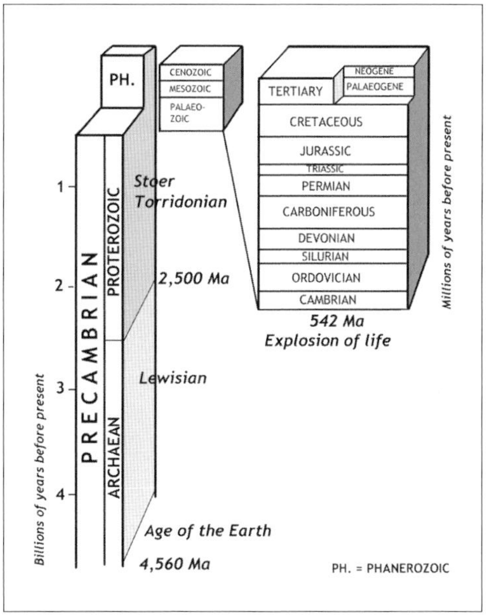

The timescale of the Precambrian is from the Earth's beginning at 4.56Ga to the start of the Phanerozoic, the start of the Cambrian, at 0.54Ga. Writers of geology books like to draw an analogy between geological time and some familiar measurement to bring home its real magnitude: the hours of the day, pages in the Bible, distance along a road, floors on a skyscraper and so on. So, being Scotland, let's take the golf course. The first 16 holes are Precambrian, abundant life is found from the 17th fairway onwards with the dinosaurs coming in at the 18th tee, lumbering along the fairway and disappearing (for the sake of golf courses, thank goodness) just before the 18th green. Humans are briefly in the hole when we finally take out the ball and return to the clubhouse. As any golfer knows, the first 16 holes count! A geological truism as well. For convenience, this huge time period of the Precambrian is simply divided into two: the Archaean, 4.5–2.5Ga and the Proterozoic, 2.5–0.54Ga, although as things stand today this division is rather arbitrary and nothing particular happened geologically at 2.5Ga to separate the older events from the younger. More recently, a commission has tried to set up new divisions, but since these are all based on absolute age data, they will probably not be used; actual rocks or location names have more relevance. At any

Figure 1.3
The Stratigraphic Column: the time library of geology. There is only detail over the last 540 million years (the Phanerozoic), since abundant complex life has existed. Through the Precambrian, from 540 million (540Ma) to 4.56 billion (4.56Ga) years, few subdivisions are used. (Timescale of Gradstein et al., 2004).

*Billion is used throughout this book to mean 1000 million.

Figure 1.4
The surface of the
Moon is 4.45 billion
(4.45Ga) years old.
The Moon's far side by
Apollo 11 (courtesy of
NASA).

rate, let us use what exists and say that these 1.2 billion-year-old (1.2Ga) Proterozoic sediments from northwest Scotland are part of the Earth's early story, a rare and valuable piece of the primitive surface.

In the village of Stoer, as in most Highland villages, the cemetery is an important place. The history of the local population is buried here, of course. But it so happens that it is also the place of the unconformity, the place of contact, between the sedimentary rocks of the Proterozoic Torridonian and the underlying crystalline, Archaean Lewisian Gneiss. This unconformity is remarkable, and for two reasons. Firstly it is because of the ages of the rocks involved; the Torridonian sediments are dated at 1.2 billion years, Britain's oldest sedimentary rocks, which is impressive enough. But the crystalline, metamorphic rocks on which they rest come from very far back indeed and are nearly 3.0 billion years old, from when continents were first being created, as described in Chapter 6: truly Britain's – even Europe's – oldest rocks.

The ancient Lewisian Gneiss mostly started as magma that cooled and crystallized and was then mangled and deformed by massive heat and pressure from its time deep in the Earth's crust, so that it is now entirely crystalline, as befits its age. The covering Torridonian, although also exceptionally old, is still 'fresh' and, like Dorian Gray, shows no marks of passing time. We expect old things to show their age – old faces are wrinkled and old rocks are twisted, and if they are not they must be possessed of some kind of magic. So it is for the Torridonian. Age is only the first amazing feature of this unconformity; the nature of it is the second. The ancient Lewisian surface, the unconformity itself, forms hills and valleys, deep crevices and vertical cliffs, all intimate features of an ancient landscape. It is still there to be seen from the cemetery, a fossil scene of huge antiquity sculpted and eroded over 1.2 billion years ago. The majestic cliffs and deep crevices in the rocks of the tiny village of Stoer have spectacularly cheated time – more than one billion years of it.

The village of Stoer itself is sited on the Torridonian sedimentary rocks, while the bare, lochan-scattered Lewisian gneiss below the unconformity stretches endlessly inland, eastwards. The layers here were originally flat but are now tilted – in geological terms, dip – at about 25° seawards, meaning that they get progressively younger westwards, away from the unconformity. For this reason the sea cliffs around the Bay of Stoer and around the village of Clachtoll (the settlement to the south of Stoer) are all Torridonian sediments. A broch established here over 2000 years ago used these Torridonian rocks, as they split easily into building blocks. Going north or south, perpendicular to the dip and geologically along strike, is to follow layers of the same age. This means that both north and south of the cemetery, we can walk along the contact between the sediments and the crystalline gneiss. If we do this, some remarkable things are to be seen.

About one kilometre south of the cemetery, along strike, just behind Clachtoll and the several houses built on the flat, cropped grass of the machair, a slab-like rock slope with grassy patches rises up inland to the east. The rock here is Lewisian and the slope, at about 25°, is the unconformity, the now exposed surface of the ancient landscape. It is expected to be flat, a long ago worn-down surface like a much-scrubbed table. At first view it does not look like much, but an experienced eye soon sees that while most of the bare rock slab forms rounded, bleached knolls typical of the Lewisian gneiss, some broad areas have a brown colour; these are Torridonian sediments. Clambering on the slope, the Torridonian is unmistakable. On exposed surfaces the rock is clearly made up of a jumbled heap of huge, angular boulders, blocks and cobbles with the spaces filled in by coarse sand and grit. A longer, careful look shows that all the boulders and large bits of rock are made of Lewisian gneiss, the rock immediately below the Torridonian, and quite a few boulders are of such fantastic, unlikely and fragile shapes that they could not have come from very far, otherwise they would have broken apart. The Lewisian, it seems, disintegrated in place, the boulders lying around like rock rubbish.

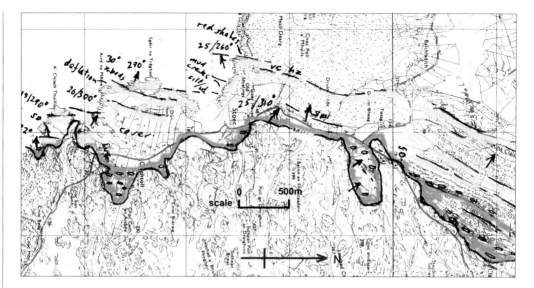

Figure 1.5
A hand-annotated
geological map of
the coast between
Clachtoll, Stoer and
Clashnessie showing
the trace of the
unconformity between
the Torridonian
sediments and the
Lewisian gneiss (base
map courtesy of the
Ordnance Survey).

There is more. The contact between the boulders and the gneiss that outlines the ancient land surface is easy to follow, the two rocks are so obviously different. Normally, a flat rock surface dipping at 25° west will trace a north–south line along strike. What happens here is that the direction actually followed by the unconformity is anything but north or south. It is more often east, towards inland and then almost west, towards the shore. Reflecting on this, it means that the unconformity surface is not flat at all, but highly irregular, and if you walk along the contact behind Clachtoll, first going abruptly one way, then equally abruptly the other, you will be tracing out the near-vertical walls of an ancient canyon in the unconformity and looking down at the blocks that fell into this narrow, 150m deep chasm (enough to hide Big Ben easily), 1200 million years ago. It is as though the dark shadows and the oppressive silence at the bottom of this incredibly old abyss are still there.

These are not the majestic cliffs and valleys of an ancient landscape that were promised. To find these we must go back to Stoer, climb the present-day road northwards through the village until just over the crest of the hill beyond, and stop where there is a turn-off across the peninsula to the lighthouse. The road now serves other residents and visitors only; the keepers are gone and a computer operates the light.

Standing on this high point, back to the village and looking northwards, is an object lesson in basic geology, a simple sight worth hours in the lecture room. The high ground is covered by ribs of rock heading north across a broad plateau to the horizon. Each rib is a continuous bed of sediment which, because it is tilted westwards, has a long dip slope but a small, abrupt cliff facing the opposite way (east), where the rock layer is cut off like a step to expose the next, older layer underneath. It looks like a shelf of big books that has just fallen over, each one where it landed, not quite on top of the one below. Settled on the edge of one of these book-steps, in the most perfect location ever seen, is the Victorian, stone-built Stoer Primary School. Beautifully alone with the rocks beside a small lochan: no trees, no roads, houses, or factories, only wild flowers and the arctic cotton. No pupil of this school could ever forget the lovely first years of their education. Regrettably, the falling number of children in the area has led to the mothballing of the school. Perhaps they will return one day, when people and governments value this landscape more.

1200 million years ago during the Torridonian, this was equally a high point, and just below it is another deep chasm like the one at Clachtoll. This one is even more dramatic, just 200m wide, cliff-sided but an amazing 250m deep, the equivalent of a sky-scraper of over 70 floors – a huge, angry cleft. Cheddar Gorge has cliffs of only 100m and Beachy Head is just 150m. Beyond is an even deeper chasm, though this one is much broader. It has steep, partly cliffed sides, is over 350m deep but beyond 2km wide, not quite a Grand Canyon, which is 6km wide and up to 1.5km deep, but a worthy monument none the less. For the moment this Stoer canyon is empty, but soon it will be filled with

Figure 1.6
1.2Ga rock rubble that collected on top of the worn-down Lewisian surface. Coast near Stoer.

Figure 1.7
The billion-year-old
unconformity between
the Lewisian and the
Torridonian was like
the Grand Canyon
(after Geikie 1886).

raging torrents surging from the northeast, bringing huge boulders hurtling down, cracking and crashing against each other, only to be left as debris in the valley bottom when the violent currents subside. Sometimes the flow is less or there may even be no water at all, but there follows again another raging torrent and rushing boulders, the violent activity and energy of a young planet. Pure billion year old imagination? No, the picture is based on scientific evidence.

The Lewisian–Torridonian contact traced on a map is a 25° slice through the ancient landscape, as this is the present dip of the originally flat layers. To draw a true depth profile of the contact, the 25° dip and the present height must be corrected for. From the simple calculations, these amazing depths emerge: this truly is a remarkable canyon. But how did it come about? We must explain how such great deep, narrow canyons could be created from hard, crystalline Lewisian rock.

On the present planet surface, such chasms are unusual and suggest quite special circumstances. The American Grand Canyon has been carved out over 10–20 million years and formed as a result of the land rising up with an ancient river still flowing through it. (Although a letter to the Highland *John O'Groat Journal* stated, '... the Grand Canyon was formed over a period around twelve months by the worldwide flood detailed in Genesis.') As the land slowly rose, the river cut down. If the rise had been too abrupt, the river would

10

have been blocked and gone elsewhere. This was towards the end of an incredibly long period of erosion that had removed many tens of kilometres of gneiss.

If we apply the rules of the present to the past, then our Stoer canyon would have had a similar history and taken a similar amount of time or longer to form. But the Grand Canyon is empty; ours is full. To fill the present Grand Canyon, sea level would have to rise nearly two kilometres or, more likely, the land would sink by the same amount. The rocks to fill it with would then have to be found. The same is of course true for the Stoer canyon. But here we see that as the canyon filled up, so the walls rotted, blocks falling into the depths and furnishing the raw material for the boulders we now find. Where these boulders are still angular and with fantastic shapes, there was no water to move them away. We can imagine a narrow crevice, too narrow for much water to flow through, filling with the rubble from the cliffed walls. Where the valleys were wider, water flowed and the blocks were rolled, worn and rounded to form the boulder heaps we find today. This must all have taken a very long time.

We can compare the Stoer canyon and crevices with the Grand Canyon or other features we find today. But this is not right. One of the most important ideas in geology, first proposed by the great geologist Charles Lyell (1797–1875), is that 'the present is the key to the past'. In other words, what we see around us today is typical forever and we can expect no difference from the past. The Grand Canyon could be created at any time, past or future, in exactly the same way as today. This is a useful idea to prevent the more excessive imaginations of what happened in the past; Hollywood could never abide by such a rule. And for the Torridonian, neither should we. It would, in fact, be an error to use today's planet as an exact model for the Torridonian; there are many important differences. The first is

Figure 1.8
A geological section through the billion-year-old canyons of Clachtoll, Stoer and Clashnessie based on ground measurements.

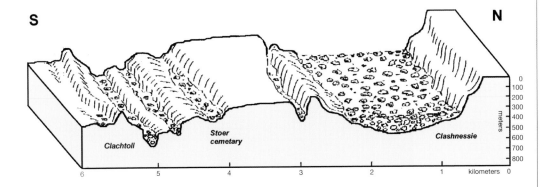

that 1200 million years ago there were no plants, the land was bare and rocks and stones covered the entire, desolate surface, like a scene from a Martian lander. So how do we dissect the rocks of Stoer to tell us about this ancient Earth?

Beyond the machair in front of the cemetery, actually the common grazings of the Stoer crofters, bare rock forms a continuous platform beneath low sea cliffs on the north side of Stoer Bay. All the beds are still tilted at about 25° to the west, which means that heading westwards along the seashore the layers being encountered become progressively younger. The first rocks on the beach, the oldest, are those that follow the remains of the violent floods. They are regular layers of red-coloured mud rocks and siltstones, fossil sediments that

Figure 1.9
This is Mars today. 1.2Ga ago, Stoer would have looked like this. Even then primitive life existed (courtesy of NASA).

formed in quiet, shallow water or open flats. They tell us that the old, scarred landscape of deep gorges and raging torrents has now been smothered into a scene of gentle maturity.

Examining these sediments in detail, layer by layer, we find features quite unlike those in the rocks below. There are no boulders or blocks, only fine fossilized muds, silts and sands; no irregular surfaces, only (originally) flat, horizontal layers. The sands show the profiles of quiet water ripples similar to those from modern lake shores and beaches. Incredibly though, there are layers with fossilized desiccation or drying-out cracks. Today we associate these with evaporated puddles or the edges of a drying reservoir in summer, but today's conditions cannot be applied to the planet's plantless past.

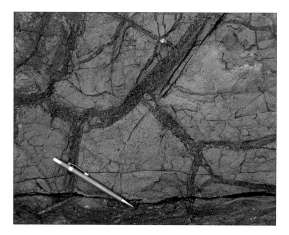

Figure 1.10
Common or garden
mud cracks –
preserved for over
1Ga. Photo from the
foreshore at Stoer.
Pencil 15cm.

The desiccation cracks occur in red silty beds 5–50cm thick. However, no matter how thick the bed, the cracks go from the top almost to the bottom, showing that the entire bed dried out at once. The cracks are wedge-shaped, wide at the top and tapering. In plan view they are the familiar hexagon. The top surface of each cracked bed is covered with light-coloured sand in fine layers, with the same sand grains filling the cracks underneath.

The explanation for these structures is fascinating. The silt beds were deposited almost instantly after heavy rain fell on nearby highlands. The rain destabilized slopes, dislocated the surface silt and washed it away as a wet, cement-like slurry, a flash flood. When the swirling, viscous slurry came to a flat area it stopped and set like cement in the place where we find it now. This was a wind-dominated climate, like present-day Mars, and the slurry quickly dried, shrivelled and cracked. The wind persisted. There are no plants and no trees so it is a hard and furious wind. Dust storms were clearly frequent and the winds so strong that only sand could remain on the ground, rolling along and falling like sand in an hour glass, into the open cracks in the now cement-hard, desiccated silts. More wind and the sand collected into the fine layered deposit we now see fossilized. The size of the grains, up to 1cm across, tells us that the winds were like those only found in the most desolate of deserts today. All the mud, dust and fine particles were consistently blown right away, only stopping by falling into water.

Desolate and wind-dried as this place must have been, there were still shallow lakes of perhaps semi-permanent standing water. The rocks provide the evidence for this. At Diabaig, 100km to the south, there are similar sediment layers to those at Stoer with similar desiccation cracks and wind-blown sands. But here, in addition, in amongst the desiccation, are the fossilized marks of rain drops, an occurrence at once seemingly so commonplace but in truth quite remarkable. It is comforting to know that rain fell 1000 million years ago. But can we tell if it was the kind of rain we would expect? To fossilize something so ordinary as the trace of a rain drop is a near miracle and takes very extraordinary circumstances. The mud must be soft and slightly wet, there must be no actual standing water, and the rain must be a short, heavy shower of large droplets, not a complete cloudburst. Once marked, the rain-pitted mud must solidify and be quickly covered by more mud or some other sediment – a very strict sequence. If any one of these events does not occur, there will be no fossilized rain print. There is an illustration in the famous book, Geikie's *Life of Murchison* written in 1875, of Old Red Sandstone fossils (about 380 million years old), one of which is called 'roots with worm burrows' but is manifestly of mud cracks with rain drop pits! So ordinary an explanation that it was overlooked. Experience shows that fossilized rain drop pits are generally produced in deserts, or as far as the Precambrian is concerned, a desert-like climate. And that is all we can really tell. Lovely as they are, these billion-year-old rain drops only tell us that rain fell then as now and that the weather was changeable.

Such fossil rain pits are generally more than 350Ma old because in beds younger than this, that sticky mud becomes marked from other sources. On the truly old rained-on surfaces we see no other marks: no insect traces, animal tracks or bird prints, no signs of life. Today, of course, such surfaces will be marked with all of these, even human footprints, but in the Proterozoic no such life existed and once formed, the rain-pitted surface remained silently untouched to be carefully fossilized, the simple event recorded for ever. But this is not quite true; all was not lifeless on the Torridonian planet. Among the desiccated slides and slurries we have just looked at there are occasionally quite thin, bright red silt intervals, which were deposited in warm, amenable, shallow lakes or lagoons. In these there was life.

It wasn't just rain that fell from the sky. Recently discovered evidence suggests a meteorite collision 1200 million years ago, during the Torridonian, which punched through the sediments and down into the underlying Lewisian Gneiss. No crater is seen today, but the material that was blown out of it raced across the surrounding surface as a ground-hugging flow of hot gases carrying molten fragments and loose sand, mud and even large, rafted masses of sediment that were near the surface. And this is the evidence we see today (Figure 1.11).

Pictures of big nuclear explosions may give you an impression of what it was like, although this was many times bigger. There are no substantial remains of the meteorite itself left; it just vaporized on impact. This was certainly the most dramatic day during the Torridonian at this location, and punctuated the normal slow build-up of sediment in rivers and lakes.

Figure 1.11
The dark-coloured rock is the ejecta material and the lighter colour are sandstone rafts carried by the flow, which are buckled and broken up.

The resulting layer of ejected material locally can be up to 15 metres thick and consists of sand, mud, and pieces of green rock, which were once molten material, and also shattered pieces of Earth's rock broken up during the impact. Chemical and microscopic mineral evidence from the layer suggests a meteorite impact and not a volcanic mudflow, which was the original interpretation. Quartz crystals in the layer have been highly stressed, a small amount of an element called iridium is found, which is an indicator of an extra-terrestrial source, and a mineral called Reidite is present in small quantities, which as far as we know is only formed with the extremes of pressure found in an impact like this. The top of the ejected layer also contains material that rained down from the sky, forming balls of fine sediment, about the size of raindrops (Figure 1.12).

As there is no sign of the crater today, some detective work is needed to try and track it down. Craters on Earth do not last long, as they erode away or get filled in, or are even caught up in great mountain building events over a longer time period. Mike Simms of National Museums Northern Ireland has looked at flow direction indicators in the ejected material and, based on an anomaly of slightly lower than expected surface gravity, has proposed that the crater, about 50km wide, was close to the village of Lairg in central

Sutherland. Gravity can be slightly lower than expected where there is a buried crater, as it is often filled in with slightly lower-density sediment than the surroundings, and can have cracks that penetrate well down into the underlying rock. The pull of gravity at any one point on the Earth depends not only on the overall mass of the Earth, but also on the mass of the immediately surrounding rock. It is not much of a variation, but can be detected with sensitive instruments.

Investigations continue as the University of Leicester is co-ordinating some further research into this type of impact ejected material. The learning journey continues for us all.

One billion years ago, if we had looked, the shallow water would have been clear and still: nothing swimming about, nothing skimming on top. But on the bottom, although nothing moves, there is life, a green slime spread all over, slightly bumpy and sometimes ruckled. It is a microbial mat, a massive colony of living, dividing, Proterozoic bacteria. As Malcolm Walter, one of the foremost researchers in the subject, writes:

> *For billions of years the shallow seafloor was carpeted with microbial mats. They spread down the continental slope and probably formed huge patches on the deep seafloor. They were widespread in lakes and lagoons. We can speculate that mats also formed extensive subaerial sheets on land. This was the scene throughout the Proterozoic, a duration of some 2 billion years.*

Figure 1.12
The green pieces in the ejected material were once molten, the red background is broken up sandstone and mudstone and the pink and black fragment is a piece of Lewisian Gneiss carried by the flow.

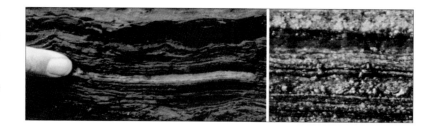

'Slime time', as someone must surely call it sooner or later. And this is what we now find in the Stoer rocks, 1.2 billion year old slime called stromatolites, fossilized as centimetre-thick bundles of very fine, slightly wavy laminations, preserved in calcium carbonate within the bright red silts. Some call these 'bacterial lawns' but gardeners will object; it may not be politically correct, but this is slime. No other word. This makes it clear that these are not childishly exciting fossils like ammonites or trilobites, and for years they were ignored (except by the notable few like Walter). Today there is a realization that even if they were slime, they were quite fundamental to the Precambrian world and their lack of looks belies an incredible influence. Hard to realize in the present world where looks are all, but without them we would not exist. Call them slime; we shall not ignore them.

Stromatolites are structures built of both organic material and sediments, although each remains separate. They are picked out in fine, calcareous-rich layers, forming, as Walter says, extensive carpets. The name stromatolite was invented to describe laminated structures found in the Palaeozoic, mainly Silurian to Ordovician (416–488Ma) sediments, which were clearly of organic reef builders and were classified as such along with the corals, although in truth the life-forms creating them were not known. An authoritative palaeontological (fossil) book from the 1950s states:

> *Various authors have suggested classification of them with algae, sponges, foraminiferal protozoans and bryozoans, but almost surely they cannot be associated with any of these.*

Clearly, at this stage there seemed to be no evidence to show what stromatolites were. Towards the end of the nineteenth century, when the same structures were first found in the Cambrian and Precambrian, it was even uncertain whether they were indeed biological or simply mineral. When Charles Doolittle Walcott (discoverer of the Burgess Shales) found a form in Cambrian limestones near Saratoga, New York State, he named it Cryptozoon, 'hidden life', since he could find no direct evidence for a biological origin. However, when he found similar forms in the Precambrian at the bottom of the Grand Canyon and then eventually in 1899 in the Precambrian of the Montana Rockies, he decided that they were indeed biological and interpreted them as built by algae. It is remarkable that, in the face of considerable evidence, it took nearly 70 years from these first discoveries for the true origin of stromatolites to be convincingly demonstrated. They are formed by a complex ecosystem of cyanobacteria and bacteria, and the reason we know this is that stromatolites are still living today.

It seems unbelievable that the one-billion-year-old fossils that we are examining should have living representatives. In fact the oldest stromatolites are dated as 3.45 billion years old from the Pilbara Supergroup in Western Australia, and similar fossils almost as old come from the Swaziland Supergroup in South Africa, the finds only being confirmed in the 1980s. Despite this fact, the cyanobacteria of these extremely old fossils can be studied living today, most famously in Shark Bay in Western Australia but also at a few other rare sites around the world, from Mexico to the Bahamas. The bacterial story they tell today is amazing; billions of years ago it was astounding.

Cyanobacteria are the principals of stromatolite growth. Commonly called blue-green algae, they are actually bacteria and are made of prokaryotic cells, the earliest and most primitive form. The cells are small, have no nucleus and reproduce by simple splitting (mitosis). There are no sexual differences. All complex life forms, ourselves included, have large eukaryotic cells with a complex nucleus and sexual reproduction which evolved out of the prokaryotic form. Cyanobacteria are recognized in the fossil record by the size of their cells. Bacteria generally have tiny cell diameters of less than 1.5μm (0.0015mm) while cyanobacteria are slightly larger with diameters over 2.5μm and up to about 60μm. Eukaryotes, however, are much bigger and for example, early single-celled algae called acritarchs have cells that measure between 60μm and 200μm, a significant difference from the prokaryotes. Cyanobacteria are photoautotrophs; that is, light is their energy source. With this energy they take in CO_2 and H_2O and create the large organic molecules they need to reproduce themselves. This process produces oxygen as a waste product that is liberated to the atmosphere. These sunlight-using microbes form the growing surface of the stromatolite. One typical microbe body shape is like a fine string of cylindrical beads, another like a light-bulb filament loosely coiled, but in each case they have one, light-sensitive, light-seeking end. In their effort to find light, the strings get tangled

Figure 1.14
A primitive prokaryote cell (left), a bacterium, compared to a much larger and more complex eukaryote cell. Complex life reproduces sexually and has eukaryote cells (schematic).

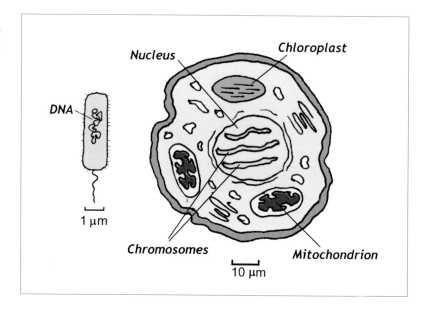

up in each other like worms in a fisherman's live bait box, and in this way build the living mat. The whole is covered in mucus, and modern stromatolites are leathery but tacky to the touch. Progressively, living forms build upon the dead and the paper-thin mat thickens. However, the cyanobacteria are not alone, and below the growing face is a population of bacteria, equally photosynthetic but which need less light than the cyanobacteria. It has been discovered, amazingly, that the photosynthetic green and purple bacteria in this undermat are so coloured as to be able to use light of a different wavelength from the cyanobacteria above. As the living mat surface builds up on dead forms, so the coloured forms themselves migrate upwards. But there is still one more community in the layers, that of bacterial forms that do not use light at all, or indeed oxygen – it kills them – but live off the dead microbes discarded by the communities above; these are the anaerobes.

When cyanobacteria first started to produce oxygen, it was not free to join the atmosphere, but was absorbed by the marine environment. Layers of oxides settled on the sea floor, with minerals such as magnetite and silica, forming what are known as banded ironstones. Only when this process was complete did free oxygen start to accumulate in the atmosphere, allowing red sedimentary rocks on land, such as the Torridonian, to form for the first time. The redness is partly produced by oxides of iron that cemented the sand grains together. This major atmospheric change is one of the few processes that have occurred only once during the life of the Earth,

Figure 1.15
Minute Cyanobacteria fossil 850 million years old from Bitter Springs, central Australia (by William Schopf).

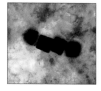

and looks irreversible; most other processes are often repeated in continuous cycles.

Finally, the environment intrudes into this living mass and the microbial mats are interlayered with ordinary sediment. From modern studies it seems that when the amount of sediment in the water is very high, for example after a storm or a flood, it swamps the growing face and covers it over. When this happens, the cyanobacteria simply search for the light, migrate upwards through the new sediment, recolonize the surface along with the other microbial forms, and start out building a mat all over again. In rough, open sea conditions this causes localized growth and 1m-diameter conical or domal forms grow up, as in Shark Bay today. In quiet waters growth is regular and flat; continuous mats result, as we find in the rocks of Stoer. In addition, cyanobacterial activity is frequently associated with calcium carbonate, and many stromatolites are preserved as limestones. Whether this means that the calcium carbonate is biological or simply a mineral precipitation is difficult to tell, but the common association of microbial mats and limestones suggests at least a biological interference in chemical affairs. This is especially so as it is proposed that the early oceans, even up to 0.8Ga, were not dominated by salt, NaCl as today, but by HCO_3- and Ca^{2+}, the so-called 'soda ocean' which made calcium carbonate precipitation easier.

So it was that Archaean and Proterozoic bacteria created reefs as large and as widespread as those of modern day corals. There are remarkable examples of stromatolites built into a stunning variety of structures as variable as any living reef, the most spectacular being tree-trunk-like mounds 30m high and with 12m relief at Dismal Lakes in northern Canada. On a pleasing practical note, the huge decorative columns of the Great Hall of the People in Tiananmen Square are actually stromatolitic limestones (it is the Hall of the State, not the people, a rather obvious political spin on words). Compared to these, our own (non-political) examples from Stoer are rather unimpressive though they, as well, are picked out in calcium carbonate. This calcium carbonate is precipitated over the surface of the growing mat due to the depletion of carbon dioxide in the water during photosynthesis.

But what does this knowledge tell us about the stromatolites of Stoer? Or equally, what does Stoer tell us about stromatolites? We can treat both questions and say that the 1.2Ga stromatolites of Stoer are a sign of Precambrian evolution and occurred at a crucial moment in their development. From Stoer it is possible to look back at the huge expanse of time before and then look forward to the amazing

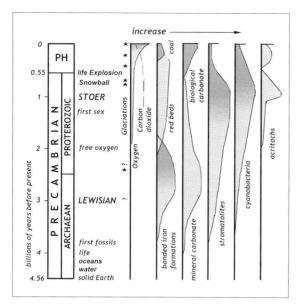

Figure 1.16
The major events
of the Precambrian
(and Phanerozoic)
showing the changes
in the atmosphere
and oceans with
parallel changes in the
development of life.

things that were to come, because evolution through the Precambrian is the evolution (or lack of it) of stromatolites and the cyanobacteria, as Malcolm Walther says. There is no other life. Evidence shows stromatolites were present very early in the Archaean, at 3.45Ga in the Pilbara and Swaziland Supergroups mentioned previously. They very gradually increased in abundance over the next 2.5 billion years: astoundingly slow progress. They reached a peak around about the time of the Stoer fossils, between 1.2Ga and 0.9Ga. After this they declined dramatically into the Phanerozoic although of course they do still exist in those rare places today, the ultimate living fossil! The fossil stromatolites at Durness on the north coast are about half the age of those at Stoer, and represent the last time they dominated any sizeable area. The only other important early Precambrian fossils are the acritarchs that are the presumed eukaryotic, single-celled algae. These too occur at Stoer, so let us examine how it was that tiny, cyanobacterial lives gave us the complexity of life today and the oxygen to support it.

The early Earth, from 4.56Ga, was fierce and fiery, as described in Chapter 6. If it rained the water boiled instantly to feed the permanent thunder-storms that raged; volcanic explosions, lava flows, dark ash clouds, steaming fumaroles and incessant meteorite impacts, all dominated. This is imagination, of course, and a rather 'Hollywood' version at that. By the time we find the first piece of real evidence, rocks 4.0Ga old, the Earth's surface had cooled, meteorite impacts were less frequent and pre-continents were beginning to form. There was a near-continuous ocean, the water being derived from the still-raging volcanoes. Even at this early time, life may have existed. The carbon found in these earliest rocks seems, to some at least, to have the tell-tale biological signature of organic carbon isotopes (explained later in the chapter). But whenever it was that it came, first life, according to lab experiments, was created in a chemical 'soup' of liquids rich in carbon and nitrogen and activated by electrical lightning discharges.

Now this really is Hollywood, even directly from the Frankenstein laboratory! First of all, I object to having been created in a vulgar soup, something posh people have as a first course before the main dish. Life deserves better than that. Not even 'broth' as suggested by more sensitive workers; broth is soup by another name. Perhaps I may accept creation in a chemical 'essence' but not a soup. And secondly, if it was that easy, did it happen just once or lots of times, and always with the same soupy result? This is actually the real philosophical question. Was the creation of life a one-off, a singularity, or did it happen often? According to *The Hitchhiker's Guide to the Galaxy* an event occurred which:

> ...*seriously traumatized a small random group of atoms [...] and made them cling together in the most extraordinary unlikely patterns. These patterns quickly learnt to copy themselves [...] That was how life began in the Universe.*

Which is as good an explanation as some. Finding life on Mars would certainly help to contribute to the real debate, and is part of the excuse for spending millions on Beagle 2, the British Martian lander which unhappily failed. Since 2004 NASA's Mars Exploration Rovers have found rocks on the surface of Mars that show ripple marks that were formed in water, similar to those you might find on the surface of a modern sandy beach on Earth. The water on Mars probably also had salts dissolved in it. Perhaps when these rocks were formed the conditions on Mars were friendlier to life than they are now. Another mission is planned for 2020 to see if microbial activity has taken place in the most likely areas, such as in the clays produced from weathered rock at the time water was present. If none is detected, we are sure the search will not end there. Fred Hoyle, being an astronomer, decided that life was extra-terrestrial, which of course simply passes the buck, avoids the soup as it were, and goes straight to the main course; life still has to come from somewhere. So the start of life is still an interesting and defiant mystery, which is definitely better than being soup.

Regardless of the kitchen, life on Earth is certain at 3.45Ga from the evidence of fossil cells of cyanobacteria from Western Australia, preserved in water-laid cherts squeezed between massive thicknesses of pillow lavas (volcanic rocks formed under water). At this same early date we find ripples caused by water, just like those at Stoer or beside the sea today; sediments and water were well established. And that is how life continued for the next 3000 million years! 'Complex' life does not appear until 0.54Ga. There are volcanic basalts and thick sequences of sedimentary rocks containing stromatolites fossilized wherever the environment is favourable. Photosynthesizing cyanobacteria are not only very, very old, they are always there. However, over this huge passage of time, if the stromatolites didn't change, the Earth around them did. The amount of oxygen produced when a cyanobacterium divides to form two cells is minute. But bacteria can produce a new generation in 20 minutes! Over billions of years, trillions upon trillions of generations, the quantity of oxygen produced became enormous, enough to cause serious pollution. This is how oxygen entered the Earth's atmosphere.

To backtrack a little. We know that the early Earth had little or no oxygen because minerals that are found enclosed in the early rocks cannot exist in today's oxygen-rich air and cannot be carried in today's oxygenated river waters, notably uraninite (uranium oxide), the source of radioactive fuel. There is more evidence than this. Huge quantities of the iron oxides, haematite (Fe_2O_3) and occasionally magnetite (Fe_3O_4), are found in finely banded layers from Archaean rocks across all the continents from Australia to Canada and South Africa. The iron in them was spewed out of volcanic vents and cracks at the bottom of the then deepening and developing oceans. This iron dissolved quickly in normal seawater but in warm, shallow seas where there was more oxygen, it was oxidized, rusted, and became converted into tiny iron oxide grains that settled out as a sediment onto the sea floor. The effect may have been seasonal because the same very thin bands can be followed for hundreds of kilometres. Banded ironstone formations, BIFs, need only a limited amount of oxygen to form, which was clearly available at the time in shallow waters. The amount of oxygen soaked up by and still enclosed in these formations is enormous, estimated at 20 times what is in today's atmosphere, and it could only have come originally from biological sources – the cyanobacteria. Water, nitrogen and CO_2 are all produced from volcanic eruptions: oxygen is not. So, bacterial slime polluted the earth with oxygen and the iron formations soaked it up for 1.5 billion years, before the equilibrium finally broke down. The volcanoes gradually ceased to erupt, the iron was no longer available, the oceans had deepened, and oxygen pollution steadily built up in the atmosphere. We are close to 2.0Ga.

At the same time as the oxygen levels in the Earth's atmosphere were increasing, the fossil record begins to change; the two events are related. Large-celled eukaryotic acritarchs, life-forms using the oxygen provided by the cyanobacteria, appear as fossils around 2.1Ga. We are now able to explain Stoer, which although nearly 1.0Ga younger, has all the characteristics of an oxygenated Earth. First of all the sediments are red, stained by iron oxide, a feature unknown before 2.0Ga because there was not enough free oxygen to fully rust the iron. Today this is a sign of hot, dry climates, but it was not necessarily the case in the Precambrian because of the lack of plant cover. For the Precambrian it is more a sign of excess oxygen and dryness, although from overall climate considerations it was probably rather warm anyway. Once introduced, the eukaryotic acritarchs, supposed single-celled algae that drifted around alone in the water, or in technical terms phytoplanktonic algae, quickly increased in both abundance and variety.

At Stoer the forms are still simple, smooth and bubble-like. Around 800Ma acritarchs reached their acme and a wide variety of forms are found; smooth, star-shaped, nobbly and with spikes on. The success of the acritarchs was paralleled by success for the stromatolites, which equally seem to have reached a peak around 1.0Ga, although for them it was success only in terms of abundance and not of variety. In fact, the rapid development of many varieties of acritarch, when put alongside the complete lack of new varieties of prokaryotic cyanobacteria, despite their numbers, is glaring. In a leap of imagination palaeontologists, like tabloid reporters, put this down to sex. Prokaryotes are non-sexual and reproduce by simple division (mitosis), so-called fission. There is no chance for error and so no evolution. Eukaryotes can be sexual and reproduce by meiosis, and in general the larger the cell size the more probable this is. The larger the cell the greater the amount of information to transfer, and once reproduction is sexual, the greater the chance of an error, of differences arising and of evolution taking place. Sex leads to mistakes. So that's not new, then. But more pleasingly, evolution is sexual. This is the present explanation for the dramatic increase in acritarch variety from about 1.0Ga. Stoer is in the midst of the sexual revolution, and so successful was it that it is still with us today. Sex it seems, despite the tabloid obsession, is not trivial and should be taken seriously. Looking back it all seems so simple: no oxygen, no eukaryotes, no sex, no evolution. What a change since then!

Suddenly, at 542Ma, there are abundant fossils of complex animals: metazoans. This is the dramatic appearance of modern life, the famous 'Cambrian Explosion'. Remarkable sites in Greenland (Sirius Passat), China (Chengjiang) and especially the Burgess Shale in

Figure 1.17
Hallucigenia sparsa,
a fanciful life form
from the mid-
Cambrian Burgess
Shale, Canada, a little
younger than the base
'Cambrian Explosion'.
Fossil (above, about
3cm long) and
(below) a possible
reconstruction
(courtesy of the
Smithsonian Museum,
Museum of Natural
History).

the Canadian Rockies, dated at between 500Ma and 530Ma (slightly younger than the explosion), have produced a huge array of wonderful and weird, beautifully preserved fossils. There are frondy seaweeds, sponges, many-legged courgette-like worms of several varieties, small trilobites similar to woodlice but quite unrelated, and other variously shaped crawlers or perhaps swimmers so weird they cannot even be described. Palaeontologists have a hard time with these fossils as they appear to defy normal body pattern rules. For example, in the case of *Hallucigenia sparsa* it is not clear whether the back spines are actually legs or the legs are really back spines, which end is which, and whether the presently identified head may actually be the back end!

There is a proposal that this was early evolution gone mad and that most of these forms were 'experiments' which did not work and so became extinct. This is of course disputed, as nature is never so wasteful, and the preservation in these *lagerstätten* (layer life) as they are geologically called, is special. Such layers provide a record of life only rarely seen, and remind us of how incomplete the fossil record is. Whatever the truth, a stunning array of varied, complex animal life is present, and it all arrived suddenly. So suddenly, in fact, that most metazoan forms of life – arthropods, brachiopods, molluscs, annelids, echinoids and even chordates (animals with back bones) – seem to have originated in the Biological Big Bang, as some call it.

For three billion years stromatolites were alone, although latterly with the acritarchs; then dramatically, complex life arrived. It was an event that worried Darwin, and still would. His evolution requires that life arrives little by little, gradually, by tiny, progressive changes, but at the Proterozoic–Cambrian boundary at 542Ma, so many sophisticated life forms arrived in a geological instant. It was as if life emerged from years of windowless, Proterozoic prison into very bright, Cambrian sunlight: the change is shocking. Modern researchers pretend not to be quite so blinded. They say that there were chinks of light under the prison door before it finally opened. What happened in the few million years before the Cambrian Explosion is crucial. During the Neoproterozoic (750–542Ma), complex life-forms did exist, but their traces are rare and their preservation unusual. In the northwest Highlands there are no sediments of this age, so let's look at the evidence from some rather remote places around the world: southeast Newfoundland, Arctic Siberia, Namibia and South Australia.

In March 1946, in the disused Ediacara mine, lost in the Flinders Ranges north of Adelaide, South Australia, geologist Reg Sprigg (or Reginald C. Sprigg as he preferred) found fossil impressions in very unpromising-looking Precambrian sandstones. Australian researchers did not at first realize that these were the earliest traces of complex life-forms. This realization only came much later when they saw a fossil called *Charnia masoni* collected by a schoolboy (called Mason) from the Precambrian rocks of Charnwood Forest in Leicestershire, England. These fossils are the first evidence of coelomic metazoans, that is, multi-celled animals with digestive systems. Some realized the commercial value of the fossils, and today they fetch a high price on the open market. New fossil locations are now kept secret. Since that first find in Australia, similar impressions of soft-bodied metazoans have been discovered worldwide, notably at Mistaken Point, a miserably foggy part of the Avalon Peninsula in southeast Newfoundland, and on the shores of the White Sea in Arctic Siberia. All the fossils come from a very limited time interval, the final 30Ma at most (+570–542Ma) of the Neoproterozoic, immediately before the Cambrian, and are collectively called the Ediacara fauna, for obvious reasons. It is not certain whether the fauna is from a completely lost world, an evolutionary cul-de-sac, or whether it actually contains early Cnidarians, such as jellyfish and worms. Some of the impressions look just like flattened jellyfish, but then a well-squashed anything would probably look much the same.

The Russian Precambrian specialist Mikhail Fedonkin found four *Dickinsonia* (a relatively common Ediacaran fossil like a well-flattened, stubby, segmented worm that can be up to 12cm long), all of exactly the same size, lined up militarily on the same slab of rock, one alongside the other. Three of Fedonkin's *Dickinsonia* were in positive relief but the last made an indent. It was a strange and very puzzling find. In a nice piece of forensic reasoning he realized that the fossilized surface had been originally covered in bacterial slime. Each time the *Dickinsonia* rested on this slime it left a trace; perhaps it grazed on the food source. This happened three times and since these are the bases of beds, left positive marks (i.e. holes in the slime). The fourth trace, leaving an impression as most Ediacaran fossils do, is the beast itself. Perhaps it got pushed by bottom currents, or maybe it moved deliberately, since the same sort of behaviour has been found fossilized elsewhere. But the example shows nicely that *Dickinsonia*, like all the Ediacaran fossils, was soft bodied and none have crossed the boundary unchanged into the Cambrian itself. The Ediacara Fauna flourished for a few million years, spread round the globe and then disappeared. Its existence, rather than helping to explain the Cambrian Explosion, only adds to the confusion and mystery.

Complex life *did* exist before the explosion; Ediacara proves it. It may have been different and soft-bodied, but it was complex compared to what had existed before, indeed very complex. So some say this means that the explosion was more of a whimper than a bang, but in reality it looks even more impressive considering the lack of survival of the Ediacara Fauna. All the forms that appear at the base of the Cambrian are new, despite Ediacara. It has often been said that the explosion was about skeletons, about animals suddenly being able to make shells from calcium carbonate or chitin or silica, and that somehow this was part of the multiplication of life forms in general. All through the Precambrian, for instance, stromatolites formed reefs and yet they were not actually using the calcium carbonate in their structure. There is a difference between a biological influence on purely mineralogical precipitation and the actual use of the mineral within the biological structure: the difference between a human skeleton and a suit of armour. Archaean and Proterozoic stomatolites were using calcium carbonate armour. Suddenly, in the Cambrian, stromatolites are found intricately preserved in calcium carbonate; they are using the mineral within their structure and, like corals, they are creating a skeleton. After 3000 million years why the change? This was clearly a big event for the stromatolites, but the biological use of minerals only occurred in 20% of Cambrian life forms. Some workers think that the development of skeletons in the Cambrian depended on phosphorous. During the Neoproterozoic there was not enough; then suddenly, perhaps due to widespread submarine volcanism, there are common phosphate ($Ca_3(PO_4)_2$) deposits worldwide at the base of the Cambrian (there is even a phosphate quarry near Inchnadamph, the subject area of the next chapter). Perhaps so much was available that it created an allergic reaction in many animals, causing them to secrete a skeleton; a bit like fast food causing obesity. No fast food, no fat people: no phosphate, no shells. So some geologists think that the Cambrian Explosion, the start of life as we know it, was caused by a vulgar chemical change. But not all.

It is sometimes the case in science that when there is an abundance of data the imagination is stifled; perhaps a case of having to spend more time understanding the information. When there are very few facts the imagination is free to wander, to be spiritual. In astronomy, creative thinking takes precedence over understanding; it is hard to get to planets. In national economics understanding dominates over being creative; money is everywhere. The problem of the Cambrian Explosion is somewhere in between; there is a lot that needs to be

Figure 1.18
Snowball Earth: at
times between 750Ma
and 575Ma, the
Earth may have been
completely frozen for
millions of years.

1. A RED EARTH

understood, but without imagination the observations don't make sense. New discoveries are made but they only add to the difficulties. Imagination, some say over-imagination, is exactly what has now come to the problem in the form of the 'Snowball Earth' hypothesis, which concerns the last 200Ma or so before the Cambrian. There are no rocks with ages between the sediments of Stoer and the Cambrian in the northwest Highlands (the Cambrian in Inchnadamph is considered in the next chapter), but remarkable evidence of Precambrian ice ages exists elsewhere. In such outlandish places as the north of Greenland, Finnmark, Svalbard, Namibia, central Australia, Death Valley, Brazil and especially Oman, there are rocks deposited by melting glaciers. The sediments are tillites and made up of a mixture of mud, sand, gravel and boulders like the debris left all over Scotland after the last ice age. These fossil tillites are from the Varangian ice ages considered to have occurred between 750Ma to 575Ma, long after Stoer but a bit before the Cambrian Explosion. And this is where the Snowball Earth idea comes in. It suggests that the two are related, that the ice ages were somehow the cause of the explosion of life. The idea is drastic; it is fun; it is fantasy; but it is also serious science.

The scientific facts are these. Glacial tillites are found worldwide in Neoproterozoic rocks with ages of between 750–575Ma. Dating these levels is not easy, so that some think there are two glacial events, some four, some even more. What is generally agreed, however, is that from the evidence of fossil magnetism (palaeomagnetism), all the deposits formed at low latitudes (close to the equator). The palaeomagnetism in the tillites is parallel to the Earth's surface as it is at the equator, and not perpendicular (vertical) to the surface as it is at the poles. In addition, the glacial deposits are very often sandwiched within a sequence of limestones below, and importantly, the 'cap carbonates' above. These are warm water deposits, found today only in places like the Bahamas and Indonesia, a match with the equatorial position interpreted from the palaeomagnetism. But how can warm water carbonates be associated with glacial deposits, which should only form near the poles? The limestones, especially the cap carbonates above the glacial tillites, are special. They have a marked, rare, negative carbon isotope composition (negative ^{13}C), which is an indication of low or no biological activity. This is very odd, as warm water limestones are associated with an abundance of tropical marine life, the life being the source of the calcium carbonate. Carbon exists in two stable isotopes: the lighter ^{12}C and the heavier ^{13}C (isotopes are atoms of the same element with different atomic weights, illustrated in more detail in Chapter 5). Animals prefer the lighter ^{12}C isotope to the heavier ^{13}C and they use it preferentially. When there is a (relatively) high amount of the lighter ^{12}C isotope left in sea water, it means that it is not being used, so biological activity must be very low. This is the case in the cap carbonates, indicating that no living things were present during deposition – seemingly impossible for a limestone. In addition, these limestones show signs of having been deposited very quickly. Ice at the equator, nothing alive, glacial tillites with warm-water limestones, rapidly deposited cap carbonates – all of this is odd, and none of it makes sense. But if you make a snowball out of the Earth it does make sense, say a few scientists. First, however, you have to make the snowball.

In the past, climate modellers used to become frustrated with their models as, when they were trying to understand glaciation, at a certain point the model would trip and the Earth would freeze completely. This never even remotely happened during the last ice ages, so it was assumed that the models were wrong. But perhaps it is this assumption that is wrong. Under certain conditions the model is apparently right, and snowballers think that those conditions existed in the Neoproterozoic. Although the details are difficult to be sure of, it is supposed that around 1.0Ga the supercontinent of Rodinia broke up and most of the bits ended up strung like a necklace around the equator; the Neoproterozoic poles were open oceans. With no land, sea ice can spread progressively towards the equator and, as the modellers found, it can cool the oceans so much that it suddenly spreads to completely freeze them. The entire Earth becomes a hard frozen snowball, which is why there are tillites at the equator.

> *At night during summer, a thin haze of glistening ice crystals forms near the white frozen sea surface; it is much colder than the atmosphere above. Most of the haze dissipates when the surface warms up during the day, although a few sparkling crystals are carried aloft by weak convection. There are no towering cumulus clouds: there is little wind. The atmosphere is far too dry for rain or snow.*

This is James CG Walker's eerie description of the Snowball Earth weather (with my excuses for changing the text just a bit). With a completely frozen surface the entire planet would have been dry and very cold with average temperatures as low as -50°C. Once frozen, the albedo of the oceanic ice prevents the sun from melting it, and it gets thicker and colder; the Earth becomes a hard, completely frozen snowball. And this is how the planet would have remained, just another white, lifeless ghost in the solar system, the sun powerless to melt the water back, but for the Earth's own, special, internal resources.

The proponents of the Snowball hypothesis were stumped when they realized that once the Earth was frozen, they had no way of melting it; the albedo of the white surface would have reflected away any warmth. Then they realized that even during the freeze-over, volcanoes would have continued to erupt CO_2 into the atmosphere. Over millions of years, the volcanic CO_2 would slowly build up a global, gaseous blanket, and because it is a greenhouse gas it could trap the Sun's heat. Despite the intense cold it is possible for the atmosphere to be slowly warmed until eventually it becomes hot enough to begin to melt the ice. Once melting began, the protective albedo disappeared and an almost instantaneous global thaw would have occurred. They had solved the melting problem.

With all the heat in the atmosphere, and the oceans open once more, huge amounts of water vapour would have been sucked from the seas and into an atmosphere now heavily charged with CO_2; the massive torrents of ensuing rain would have been strongly acid (with carbonic acid, H_2CO_3). The heavy, persistent, acid rain fell on a land which had been ground down and pulverized by millions of years of glacial action, so that the now powdered soil easily reacted with the acid rain to be swept into the oceans and rapidly deposited to form the cap carbonates. This explains why the cap carbonates immediately follow the glacial deposits, why they were so rapidly deposited, and why they have the marked negative $\delta^{13}C$ anomaly. An explanation so neat and so scientifically clever that it leaves the sceptics stunned and silent, like the execution of a brilliant and unexpected checkmate. It underpins the whole snowball theory.

Clever though it may be, the idea has problems, an important one being that if the Earth froze over for millions of years, how did any life survive? Although the existing life forms were simple-celled bacteria and are found in some quite extreme environments today, such as hot springs in Iceland and deep rocks in oil wells, even they could not survive such extreme conditions for so long. Also, the forms that live in extreme environments have unusual forms of metabolism; the forms that survived the snowball were normal light-users. Somewhere, there had to be light. Volcanoes come to the rescue once more. In the few, isolated locations where there was volcanic warmth, so it is proposed, there could have been open water where life did survive. It does not need much to support a large

colony of tiny, primitive cells. Possibly, because of the extreme and changeable conditions, surviving life could have been forced to begin to adapt. Since populations would have been isolated, their adaptive solutions would have been different. When melting eventually took place and life spread anew though the oceans, the new varieties would have intermixed, and this could have led to the development of complex life forms where cells performed different tasks but were interdependent within one colony. Not only can the problem of life's survival be explained but it also offers a possible explanation for the beginning of complexity.

Many of these ideas are new and in need of acceptable proof. There are as many opponents as there are proponents and even more geologists, palaeontologists and biologists or simply non-specialists waiting for good evidence from the rocks. It is nice to say that volcanoes will provide life-preserving heat but most are like fickle lovers and run both hot and cold, they have long periods of extinction and inevitable cold. There is no proof that they acted as survival

Figure 1.19
Torridonian sediments
form the 'Split
Rock', the logo of
the Assynt Crofters'
Trust (courtesy
of Hugh Webster,
Stockscotland).

shelters. The various environments that could have forced adaptation probably existed before Snowball and there is no observation of independently viable cells combining to form a single life-form. And if the Earth was entirely frozen, how come that in some locations, like Death Valley, there are huge thicknesses of tillites, sediments which require ice movement and alternate freezing and melting? 300m of tillites did not form in a geological instant. But the events around the Neoproterozoic–Cambrian boundary and the dawn of life cannot be understood without imagination, since the facts alone are confusing. Carl Sagan was famously wont to say that 'extraordinary claims require extraordinary proof', and this is what the Snowball theory seems to provide. At present it suffers from its own popularity. Calling it the catchy 'Snowball Earth Theory' has made it well known, but the celebrity has spawned a whole journalistic vocabulary around it. A contending theory is called slushball; one critic titled his paper 'Not a snowball in hell's chance': a nice book on the subject has a chapter called 'Snowbralls', and so on. It makes it sound as though science is not involved, and is indeed indicative of some of the people and the emotions involved. 'Dogma is dominating data', one observer writes. Real science has now got something to work with and needs to get back to its slow, careful, scientific business: no press releases, no scientific spin and no sound bites. It is not such fun without them though.

This chapter set out to look at the amazing and unexpected record of the early Earth to be found in a remote corner of the Scottish northwest Highlands, a world which is generally hidden and most often ignored. Yet with a little effort, the rocks of Stoer have led us into their intimate world: treeless, raw winds, hurtling torrents and deep canyons, a meteorite impact, heavy rain drops and slimy puddles. And what slime. We have speculated about the origins of life, the Cambrian Explosion and ancient global glaciations or Snowball Earth. Presently, all we can do is go back to where we began, to a late spring evening and a sun setting over the Bay of Stoer. The beauty of Cul Beag, Cul Mor, Suilven, Canisp and Quinag is quite undiminished by some knowledge of their rocks. Those shadows of a late spring evening on Suilven are perhaps less dark than before, but best enjoy the light rather than dwell on the shade. And homage to John Macculloch.

Further Reading

Books, Pamphlets

Ager, Derek. 1981. *The nature of the Stratigraphical Record*. London: Macmillan Press. 2nd Edition, pp.122. ISBN 0 333 31077 2

Bengtson, Stefan (Ed.) 1994. *Early Life on Earth*. Nobel Symposium No 84. New York: Columbia University Press. pp.630. ISBN 0 231 08088 3

Craig, G.Y. (Ed.) 1991. *Geology of Scotland*. London: Geological Society London. 3rd Edition, pp.612. ISBN 0 903317 64 8

Gould, Stephen Jay. 1990. *Wonderful Life*. Hutchinson Radius. pp.347. ISBN 0 09 174271 4

Lamb, Simon and Sington, David. 1998. *Earth Story*. London: BBC Books. pp.223. ISBN 0 563 38799 8

Levin, Harold L. 1996. *The Earth Through Time*. Philadelphia: Saunders College Publishing. 5th Edition, pp.568. ISBN 0 03 023751 3

MacAskill, J. 1999. *We Have Won the Land*. Stornoway: Acair Scotland. pp.224. ISBN 0 86152 221 4

Macculloch, J. 1819 *A description of the Western Islands of Scotland, including the Isle of Man, comprising an account of their geological structure, with remarks on their agriculture, scenery, and antiques.* 3 Vols. London.

Morris, Simon Conway. 1998. *The Crucible of Creation*. Oxford: Oxford University Press. pp.242. ISBN 0 19 850256 7

Schopf, J.William. (Ed.) 1992. *Major Events in the History of Life*. Boston: Jones and Bartlett. pp.575. ISBN 0 86720 268 8

Schopf, J. William.1999. *Cradle of Life*. Princeton, New Jersey: Princeton University Press. pp.367. ISBN 0 691 00230 4

Walker, Gabrielle. 2003. *Snowball Earth*. London: Bloomsbury. pp.269. ISBN 0 7475 6433 7

Further Reading (continued from page 35)

Scientific Papers

Amor, K., Hesselbo, S., Porcelli, D., Thackrey, S., Parnell, J. 2008.
A Precambrian proximal ejector blanket from Scotland. *Geology*, April 2008

Barber, A.J., Beach, A., Park, R.G., Tarney, J. and Stewart, A.D. 1978.
The Lewisian and Torridonian Rocks of North-West Scotland. *The Geologists'
Association*. No. 21, pp.99.

Brasier, M.D. 1992. Introduction, Background to the Cambrian Explosion
(thematic set). *Journal of the Geol. Soc. London.* Vol.149, pt. 4, July.
pp.585–655.

Flemming, G. C. 1989. Petroleum Geology of North Greenland. *Grønlands
Geologiske Undersøgelse.* No.158.

Hambury, M.J., Fairchild, I.J., Glover B.W., Stewart, A.D., Treagus, J.E.
and Winchester, J.A. 1991. The Late Precambrian Geology of the Scottish
Highlands and Islands. *The Geologists' Association*, No 44, pp. 130.

Hoffman, P.E. and Schrag, D.P. 2000. Snowball Earth. *Scientific American.*
Vol. 282, No. 1, p. 5–57.

Hoffman, P.E., Kaufman, A.J., Halverson, G.P. and Schrag, D.P. 1998.
A Neoproterozoic Snowball Earth. *Science.* Vol. 281, pp.1342–1346.

Parnell, J., Mark, D., Fallick, A., Boyce, A., Thackrey, S. 2017. The age
of Mesoproterozoic Stoer Group sedimentary and impact deposits, NW
Scotland. *Journal of the Geological Society, London.* Vol 168, pp.349–358.

Simms, M.J. 2015. The Stac Fada impact ejector deposit and the Lairg
Gravity Low: evidence for a buried Precambrian impact crater in Scotland?
Proceedings of the Geologists' Association. 126, pp.742–761.

Stewart, A.D. 2002. *The Late Proterozoic Torridonian Rocks of Scotland:
their Sedimentology, Geochemistry and Origin.* Geological Society London,
Memoir No. 24, pp.130.

Walker, James C.G. 2001. Strange Weather on Snowball Earth. *Earth System Processes*. Conference, Edinburgh. June 24–28. Abstract. p.101.

Walter, M.R. 1994. Stromatolites: the main geological source of information on the evolution of the early benthos. In: Early Life on Earth. Nobel Symposium No 84. Bengston, Stefan (Ed.), New York: Columbia University Press. pp.270–286.

Chapter 2   DEEP SCAR: THE MOINE THRUST

The story of a bitter scientific controversy

AHighland battle it was, but there was no killing, no fighting, only the late night scratching of pens, the writing of acerbic letters, pointed gossip and the occasional emotionally charged lecture theatre confrontation. The dispute was no less bitter for that, although from it was to come one of the most significant discoveries in Scottish geology. The controversy was over the fundamental understanding of the geology of the entire northwest Highlands – perhaps we should say misunderstanding – for this is what it was for some 60 years. The scientific problem became evident in the 1820s, and could have been solved by the middle of the century but for what has become known as the 'Highlands Controversy'. This prolonged it until almost the century's end, fuelled by all the human failings of social and academic status, the personal possession attached to scientific ideas, and the ideology inherent in building the British Empire. In the waging of arguments, science was compromised, and when the science was compromised so were those arguing. In the end, of course the science did recover, but reputations still remain stained. Despite being about the Highlands of Scotland, much of the dispute took place in London. Scottish affairs are ever thus. It was started by nineteenth-century gentlemen scientists with money and big egos, but was finally settled by scientific, hardworking professional geologists, paid for by the government: by civil servants. It is a story of scientific progress overtaking its older practitioners, a story very familiar today in the computer revolution, but also a story of ethics in science, equally pertinent today.

The Highlands Controversy began in the mid-1800s and concerns especially the area around Assynt. It is about the Moine Thrust.

The last area (as ever) of the British Isles to be geologically surveyed was the Highlands, ending finally with the northwest of Sutherland. Knowledge of the area, as we have seen with the Precambrian, was gained from the coasts inwards,

Figure 2.1
The dark waters of Loch Assynt in a weak winter sun. Assynt was the site of a bitter geological controversy in the 19th century.

geologists working in the way a caterpillar eats a leaf – edge first. Practically nothing was known before the visits of John Macculloch and it was he, either prospecting for limestone or working for the Trigonometrical Survey, who really began the serious work in the early 1800s. Talented as he was, Macculloch made some errors of observation and interpretation, as is normal and to be expected in any science advancing as quickly as geology was in these years. It was coming out of its biblical and speculative phase into a more modern scientific one based on observation and logical inference. However, what really caused the controversy was that Macculloch's wrong interpretations were wholeheartedly taken on by Sir Roderick Impey Murchison (1792–1871), one of the geological 'Titans' of the 1800s, who took possession of the ideas and used them as his own, perhaps innocently, perhaps wilfully. He persuaded many reputable geologists

of his time that his, actually Macculloch's, interpretations were right (although they were not) and such was his reputation that the geological world remained convinced even for 20 years after his death. Murchison was the first to identify and define the Silurian System, initially in Wales; but eventually he made it into his own, personal empire, established in the wake of the parallel British Empire and in almost as many countries. Being a Highlander by birth, Murchison's final dream was to make the Highlands part of his empire; to make the Highlands Silurian. It was a dream he forcefully realized.

Let us first establish what was understood of northwest Highland geology around about 1850, when the Highlands Controversy began to affect events. In 1819 John Macculloch had published a geological account of Scotland from his mainly coastal observations. The account included a stratigraphy and structure for the entire Northern Highlands illustrated by geological sections. At the time, rocks were still being identified in terms of the 'Universal Sequence' of Abraham Werner and the Neptunists (Chapter 4), and created in order of Primary, Secondary and Tertiary from the original, primeval ocean. Primary meant crystalline and devoid of fossils, Secondary, slightly tilted and with fossils not familiar today, and Tertiary, unmoved and with easily recognized fossils. So at the time, rocks were generally properly classified as, for example, limestone, gneiss and quartz rock, but in stratigraphic terms were conceptually grouped into Primary, Secondary and so on.

The mountains of Sutherland, the central mass of the county, of gneisses and schists, were Primary, being crystalline and devoid of fossils. So also was a large area of low land along the west coast and over the Outer Isles (i.e. today's Lewisian). On the east side of the Primary mountain mass, over Caithness and east Sutherland, the

Figure 2.2
A geological section along the North Sutherland coast from Durness to Loch Hope, drawn by John Macculloch in 1819. Limestone, Gneiss and Quartz Rock are interbedded and dip regularly to the west (after Macculloch 1819).

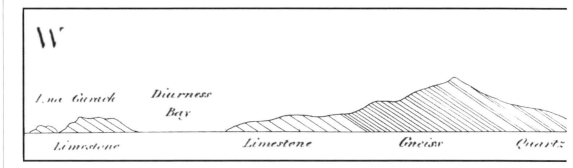

crystalline rocks were clearly, unconformably covered by fossiliferous Old Red Sandstone sediments, classified as Secondary under the Werner system. On the west side of the mountains, in northwest Sutherland and Wester Ross, the red sandstones (today's Torridonian, Chapter 1) were thought, reasonably enough, to be Old Red Sandstone as well, and similar to those of the east. In addition, a band of Quartz Rock (quartzite), and occasional limestones (today's Cambrian), were recognized. Interestingly, Macculloch indicated organic remains in the Quartz Rock, the pipes of the so-named Pipe Rock.

The geological sections drawn by Macculloch show a regular series of beds of gneiss, Quartz Rock and limestone in various repetitions, dipping eastwards under the main Primary, crystalline Sutherland mountain mass. Such a section was less surprising then than it is to us now. The presence of crystalline rocks interbedded with typical surface sediments was possible for the followers of Werner, who considered that gneiss and schist were deposited from the hot, primeval, shoreless ocean and not formed deep in the Earth. But on his sections Macculloch does not say whether the beds were Secondary or Primary; he made no clear decision. So in summary, the rocks of the northern Highlands were made up of a central Primary crystalline mountain mass framed on the east and on the west by gently dipping, Secondary red sandstones, which contained fossils in the east. Quartz Rock and limestones in the west seemed to dip under the central Primary mountains and be interbedded with them.

Visiting the area today and looking at the locations of Macculloch's sections, it is obvious that they are both a sketch and a hopeful attempt to explain something extremely difficult to unravel. They create a simple interpretation out of impossible complexity and indeed were an excellent effort, though in hindsight he missed essential elements. Once

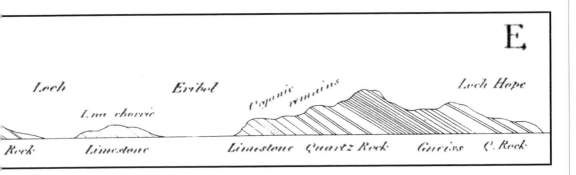

Diurness bay to Loch Hope.

Figure 2.3
Sir Roderick Impey
Murchison (1792–
1871), one of the last
geological 'Titans'
(from a picture by
H.W. Pickersgill, RA,
photo courtesy of
British Geological
Survey).

proposed, Macculloch's explanations and observations seem to have been resistant to change, and even in 1855, 20 years after his death (in 1835), his simplistic documents and posthumous geological map of 1836, were still willingly accepted by Murchison and others as quite satisfactory. The first reasonable explanation of any new scientific phenomenon always lasts longer than it should, for simply by existing it gains credibility. The ideas and observations had seemed reasonable enough to Murchison in his early years when he was too inexperienced to gainsay them; then in his later years it suited him to accept them and he made them his own. This was only one reason for the Highlands Controversy. Another, which is still true today, was due to the sheer remoteness and ruggedness of the area: few geologists had ever actually seen the rocks that were being argued about. The two aspects combined set the theatre for political manoeuvering, social slight and professional errors, mainly caused by Murchison's dominant position and obsessive character, and eventually the self-same errors from his protégé Archibald Geikie. Murchison himself was to find allies for his dispute in the Geological Society of London and perhaps even more so in the Government Geological Survey of England and Scotland, while his first, harassed opponent was simply a lone professor, James Nicol, at the University of Aberdeen, a long way from London and of no social importance. Eventually, and only eventually, opposition was to come from a solid array of scientists and Murchison's declarations were completely discredited.

In the early to mid-1800s, at the time of the Highlands Controversy, when a series of layered rocks was observed, it was not realized, reasonably enough, that layering may not be just younger over older. In normal sedimentary successions this is the case, of course; the younger layers lie on the top of older in accordance with the law of superposition, one of the paramount geological principles. But we now know that in certain situations, Earth forces can slice up rock layers and then force them into a new sequence, so putting older strata over younger. When this happens, however, there is evidence of these forces, and older layers above will be separated from the younger layers below by surfaces of grinding, breaking and annealing: in geological terms, by thrust faults. But in their work, Macculloch, and Murchison after him found no difficulty in drawing geological sections in which sediments (actually younger) were found below crystalline rocks (actually much older) in a natural succession. And this is where the Highlands Controversy became focused – on the apparently normal succession of sedimentary followed by crystalline rocks. But let us look at the modern explanation of northwest Highland geology before we look again at the mistakes that were made in the past. It is more fun feeling superior to important historical figures than making the same mistakes as they did.

The long black waters of Loch Assynt fill a deep glacial scour that heads westwards to the sea. Standing outside the Inchnadamph Hotel at the head of the loch and looking down over the cold, cheerless waters, a small mound above the lochside with a neat cairn on top can be seen in the foreground. It commemorates two geologists. Monuments are public things, meant to be sufficiently impressive to remind us all of some event or some worthy who should receive our attention, an old-style publicity exercise. That the misguided Duke of Sutherland has a massive, 30m statue glaring down on the village of Golspie that he helped to fill with his own evicted tenants from the east Sutherland clearances of the 1800s, is an indication of what statues are generally about. His victims were even asked to pay for it!

The Duke of Sutherland still stands. But this monument in Assynt is quite different. With its isolated location, it can only possibly be a gentle reminder to those who come this far north and who probably know what it is already. It is in absolutely no way domineering. The cairn is dedicated to Benjamin Neeve Peach (1842–1926) and John Horne (1848–1928), who worked for the Scottish branch of the government Geological Survey of Great Britain. Peach was a Caithness

Figure 2.4
A simplified section across the Moine Thrust near Assynt. The much older metamorphosed Moine schists, which make up the main Highland mass, are separated from the younger Cambrian limestones by the Thrust (modified from Lake and Rastall, 1927).

DIAGRAMMATIC SECTION ACROSS THE NORTH-WEST HIGHLANDS.

L, Lewisian gneiss ; T, Torridon sandstone ; a, Arenaceous series of Cambrian ; b, Middle series ; c, Calcareous series ; M, Moine   t, Thrust-plane.

Note.–The overthrusting is much more complex than is shown in this diagram.

Scot (his father had also been a keen geologist) and Horne a Stirling-born Scot. They were legendary, life-long friends. In photographs Peach is large and heavy with a mop of fair hair, Horne thin and carefully dressed having a favourite bowler hat for the field. Their fame and this monument came from their superb geological work in explaining, to everyone's amazement and satisfaction, the mysteries of this remote northwest corner of Sutherland, which in turn led to a significant advance of geology worldwide. The words on the cairn read

> *To Ben N. Peach and John Horne who played the foremost part in unravelling the geological structure of the North-West Highlands. 1883–1897. An international tribute erected in 1930.*

Murchison was knighted in his lifetime to his obvious, one might say smug, satisfaction but he has no such perfectly located permanent monument to peer recognition.

Assynt is geological Mecca and an obligatory visit for the complete geologist. It is soul-finding country. Many young students come to look at this classic area, perhaps to visit the cairn, but certainly to map out the rocks for themselves, probably wondering how the old geologists managed to do so well. They often stay at the Inchnadamph

Figure 2.5
John Horne (left) and Benjamin Peach in a now classic photo, on a bench outside the Inchnadamph Hotel in 1912 (courtesy of British Geological Survey).

Field Centre, an old estate house recently renovated, appropriately, by a geologist. From there they can look the short distance across the stony Traligill burn to the white-painted Inchnadamph Hotel where Peach and Horne spent so much time while they were doing their geological research.

The magnificent work of the two friends, along with Gunn, Clough, Hinxman and Teall, was published in a Memoir of the Geological Survey of Great Britain in 1907. It took a massive 14 years to complete! Copies are usually leather-bound, printed in Glasgow, very thick, and contain hand-coloured maps as well as some of those brownish, strangely contrasty, difficult to make out, early photographs. In the margins of the well-thumbed pages there are often hand-written notes left by the many readers. This thick memoir has certainly been read and re-read over more than the 100 years since it was written, and it is still often missing from the library when needed for reference. To begin to understand the geology of the northwest Highlands you can spread open the modern geological map produced by the British Geological Survey or you can open the hand-coloured Peach and Horne map of 1907, equally from the Survey. There is little difference between the two, and the old map is preferable. It suits the rocks better. Either map shows the dramatic line of the Moine Thrust, marked as a belt of complex colours, starting out on the north coast at Loch Eriboll and going southwest, passing through Assynt, continuing just east of Ullapool, by the head of Loch Maree, and finally skimming the southern end of the Isle of Skye before disappearing out into the Irish Sea. Through the Assynt area the band broadens up to 11km wide between Loch Glencoul and Leadmore Junction, but everywhere else it is only 1.6km wide despite being over 160km long.

Within this narrow band, the geological structure is difficult and complex. This is why it took so long to understand. In it is the clue to the whole structure and age of the huge mass of the Highlands, not just northwest Sutherland. To the left of the band (west) are the old rocks of the Lewisian which stretch even beyond the Outer Isles. On them are patches of the red sands of the Precambrian Torridonian described in Chapter 1. To the right of the band (east) is the great area of the metamorphic Moines, deposited in the late Precambrian and possibly partially equivalent in age to the Torridonian, forming the main mass of the Highland mountains. Between these, Cambrian to Ordovician limestones, Lewisian gneiss, Torridonian sandstones and various igneous rocks are sandwiched together in apparently arbitrary order, although all are tilted moderately to the east-southeast, as

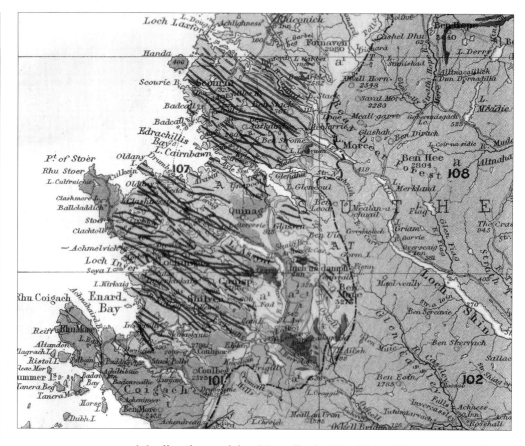

Figure 2.6
The geological map of the Assynt Window attached to *The Geology and Scenery of Sutherland* by Henry Cadell, 1896, based on the work of Peach and Horne. Light pink = Lewisian, Mauve = Moine, Brown = Torridonian, Blue and Yellow = Cambrian, Red = granite and other igneous rocks.

originally observed by Macculloch. The Moine Thrust proper forms the top of the sandwich, that is to the east of the coloured band. Other less significant thrust planes cut through the rocks below, the lowest generally being called the Sole Thrust, mainly on the west margin of the colours. Even modern, simplified cross sections of the Moine Thrust are complicated, so we shall start from simple principles.

Thrust faulting is actually quite common, but only along the outer flanks of forming mountain chains. The thick sedimentary rock piles that mountains are made of are squeezed together between rigid, converging plates of the Earth's crust. As they are squeezed, the rocks rise upwards and move outwards, like wet cement between two bricks. Since geological events take millions of years, the sheets of rock move horizontally for hundreds of miles, but very, very slowly, away from the mountain core. As the great sheets of rock move outwards, the surface on which they slide becomes entirely changed. There is too much pressure and heat for them to become actually ground up but they become physically modified and severely tectonized. The individual crystals which make up the rock on the surface of movement are bent then healed, then bent again and healed many, many times, to become what is geologically termed a mylonite, a rock so modified that it has been completely reformed. It never melted, but does produce a layer that creeps. Above the mylonite, strangely, the moving sheet of rock stays mainly intact. Such thrust faulting is seen as an integral part of mountain-building, and in the Himalayas, for instance, which are being squeezed between rigid India to the south and rigid China to the north, lines of thrust faults occur all along the mountain flanks, especially to the south overlooking the Ganges plain.

Sheets of rock bounded by thrust faults are typically many miles long, tens of miles wide and only a few thousand, even a few hundred, feet thick. They have been much explored by oil companies all over the world as they often contain oil and gas. It is hard to believe that such thin sheets of rock can remain in one piece and be pushed like this for so many miles, but engineers know it to be the case. The Dornoch Firth Bridge in east Sutherland was built entirely from the south shore of the Firth close to the Glenmorangie whisky distillery and pushed out, section by section, to the north shore 1.6km away, sliding on large Teflon plates atop the preconstructed legs. Each week a new 7.5m long section of concrete causeway, reinforced with steel bars, would be built on to the rear of the existing length of causeway. Every Monday morning, huge rams would be placed behind the new, now dry and hardened section, and the entire increased length, thrust out further over the water. Watching from the north shore, the causeway silently got closer and closer. After two years, the rams pushed nearly a mile of bridge to finally make the northern landfall: engineering imitating thrusting. The Moine Thrust moved more than 80km), but it took several million years. Interestingly, this means the Highlands only moved 1cm a year, a lot slower than the 0.8km a year of the Dornoch Firth Bridge.

The Moine Thrust is the evidence left of the north-westwards movement of the Sutherland Highland mountain mass over the old, stationary Precambrian sediments and Lewisian gneiss, the so-called foreland. The movement on the fault probably occurred during the late Ordovician to early Silurian (455–430Ma). The dates are now being further refined by isotope dating of crystal formation within the mylonite. This agrees with the fact that the uppermost Durness Limestones, Ordovician in age, are affected by the thrusting, which gives an oldest age, while the crystalline Moines are unconformably covered by the Devonian over eastern Sutherland and Caithness to give a youngest age. We are seeing a structure that was once part of the mountain's roots and buried below a huge thickness of rock, but thanks to that beautiful work of Ben Peach and John Horne, is now seen to be typical of mountain-building worldwide. Let us look in more detail at their work before going back to the controversy that provoked it.

Peach and Horne spent a great deal of time working in the Assynt area – hence their familiarity with the Inchnadamph Hotel! It is here that they undoubtedly found their inspiration, and the area still retains the geological key. In what geologists call the 'Assynt Window' several thrust sheets are piled one on top of the other. It is this that causes the band of colour to expand to over 10km instead of the usual one kilometre or so elsewhere, and for the mountains to rise to their highest in Sutherland, to the peaks of Conival at 987m (3240ft) and Ben More Assynt at 998m (3273ft). The mountains are so unusually high for the area that during a winter blizzard in 1941 an RAF training flight crashed into them, killing all six on board. Two half-buried piston engines, scattered debris and a simple war grave still mark the spot. The local Air Training Corps Unit still tends this grave. To understand the geology of the area and to determine the normal order of the rocks affected by the Moine Thrust, it is instructive to make an eastwards traverse along the shores of Loch Assynt towards the hotel and nearby Stronchrubie cliff. Then the geology of the heights of Conival and Ben More Assynt can be compared to the much lower, although equally important, Knockan Crag, a little way to the south.

At the start of the traverse along Loch Assynt shore, Lewisian Gneiss outcrops in the low, rolling, lochan-covered ground 5km or so west of Inchnadamph. In the loch, just offshore are irregular, tree-covered islets of Lewisian, and if it wasn't for the destructive feeding habits of the sheep and deer, especially the deer, the whole shoreline would be tree-covered as well. Heading east along the lochshore road, the Lewisian unconformity here is seen to be knife-sharp but undulating around road level for over one kilometre. Each

Figure 2.7
The tree-covered islands in Loch Assynt that featured on the second-class stamps issued in July 2003 are Lewisian Gneiss mounds within the Torridonian unconformity. View west of Skiag Bridge.

of the offshore islands, a roadside quarry and several road verges show red-brown, shaly, thin-bedded Precambrian Torridonian, banked up against the smoothed but lumpy surface of the underlying gneiss. This is the same unconformity described in Chapter One. and represents a time gap of up to two billion years, which is nearly half the age of the Earth.

Because the beds are tilted gently eastwards, the unconformity eventually disappears into the loch waters and the younger, overlying Torridonian rocks are seen. Looking away from the loch to the north provides a dramatic view of these red-brown, Torridonian sandstones that rise to the crests of Quinag, 810m (2650ft) overhead. The sandstone strata that give the hill its characteristic, stepped profile look generally horizontal but in fact also dip very gently at around 5° to the east, the height of the crests proving that the sand sequence can be at least 810m thick.

Less than 1.6km along this east-progressing route, the Torridonian sandstone steps are suddenly replaced at loch level by an abrupt cliff of brilliant white rock. The pile of rock that is Quinag has remarkably disappeared, for these white rocks are the basal Cambrian Quartzites, the first rocks of a new era and containing signs of the 'explosion of life'. At the base of the white quartzite another big unconformity has been crossed and more time lost: about 350 million years, the top of the Torridonian being dated at 0.9Ga and the base of the Cambrian at 0.54Ga. The Cambrian layers dip at 15° to the east, and from a sketch of the rocks like the geological section shown in Figure 2.4, it is possible to see how the huge pile of Quinag has been completely eroded away down to the white quartzites.

The brilliant white, hard, brittle Cambrian Quartzites (the Quartz Rock of the older authors) are beautifully exposed in today's road cuttings and show cross-bedding structures thought to indicate sediments moved by tidal currents. A marked ridge formed by the hard quartzites can be seen heading off up the east slope of Quinag, away from the loch, as a result of the angular unconformity over the Precambrian Torridonian beneath. The contact itself is covered by bog at loch level but is exposed in streams on the hillside above where there is a thin pebbly horizon, a sign of a surface having been worn down by erosion. These hard, distinctive quartzites can be found again further along the traverse.

The road slides down a dip slope within the basal quartzites to Skiag Bridge. These days it is more road junction than bridge, which is hardly noticeable. Near Dornoch there is a prominent road sign saying 'Littleferry (no ferry)': it was stopped 60 years ago, but local habits in the Highlands are hard to change. In the cutting just at the junction and along the lochside road are excellent examples of the famous Pipe Rock: quartzites packed with vertical 'worm' burrows, here naturally picked out against the darker reddish background in light colours. The pipes are remarkable because they are part of the Cambrian

Figure 2.8
The outcrop of Pipe
Rock at Skiag Bridge.
Crowded vertical
'worm' burrows, part
of the 'Cambrian
Explosion' of life. Scale
in centimetres.

Explosion described in the previous chapter. These rocks show that worm-like animals had already evolved by the earliest Cambrian, 540 million years ago, and that there was sufficient biological food available in the shallow seas to support them in huge numbers. They also show that predators were common enough to require the worms to hide themselves in deep burrows to survive. In fact signs of abundant ancient life are evident all along the following half-mile of road, through more Pipe Rock and then the following brownish, thin-layered, sometimes shaly, Fucoid Beds. In these the burrows are mainly horizontal as the worms, or other more complicated animals, wriggled over, or just below, the sea bottom instead of burrowing into it; and indeed, there are also rare trilobite remains. These horizontal burrows were originally thought to be fossil seaweed stems, so the beds were named 'fucoid' beds. At the top of the Fucoid Beds is a thin, distinctive layer, the Salterella Grit, named after the small curved fossil tubes of the worm *Salterella*, which can be extremely abundant on some surfaces. The Cambrian Explosion was clearly real.

Ardvreck Castle is built on the limestones that immediately succeed the Salterella beds. The castle is the long-ruined home of the MacLeods of Assynt and has recently been saved from falling into yet another lost piece of Highland history by the local heritage society. From this point onwards the succession consists of dark grey GhrudaidhLimestones, which here are affected by thrust tectonics and often associated with dark, igneous sills. The Sole Thrust is close to the base of the succession.

Nobbly, brecciated limestones continue in the hillside overlooking the ruins of Calda House (built by the Mackenzies in 1695 after they had massacred the castle-owning MacLeods). They are cut by numerous minor thrust planes that are injected by grey, igneous sills. In the slopes above, by Achmore farm, there are very visible thrust folds, which are much mapped by visiting students. Down at road level and from here back to the Inchnadamph Hotel, the dark Ghrudaidh Limestones continue to be rubbly and much affected by the thrusting. It is only once back at the hotel that the rest of the normal limestone succession can be seen by looking south towards the imposing Stronchrubie cliff. The dark Ghrudaidh Limestones form the steep, grassy slopes at the lower part of the cliff, and the light grey Eilean Dubh Limestones (the youngest beds of the succession here) form the topmost vertical face of this 120m-high feature.

It took a while for the old geologists to establish and to agree upon this complete, normal succession of rocks. That is, stratigraphically upwards: Lewisian Gneiss; red-brown Precambrian Torridonian sandstone; a Cambrian succession of Basal Quartzite, Pipe Rock, Fucoid Beds, Salterella Grit, Ghrudaidh Limestones and Eilean Dubh Limestones. The Lewisian and the Torridonian are separated by a major irregular-shaped unconformity that represents a buried landscape. The Torridonian and the Cambrian are separated by a second major unconformity cut very flat by a sea eroding the Torridonian and Lewisian. At the top of the Cambrian limestones, not yet encountered, is the Moine Thrust.

We can now divide our attention between the heights of Conival and Ben More Assynt, and Knockan Crag, 16km to the south, just beyond the village of Elphin on the road to Ullapool. The Assynt Window needs to be explored first before looking back into the 'window' from the outside.

The track from Inchnadamph to Conival and Ben More Assynt runs up Gleann Dubh alongside the Traligill river to where the major tributary Allt Poll an Droighinn runs into it. Just beyond, at the white-painted Glenbain shepherd's cottage, the track peters out into a mountain path, crosses a grey limestone karst pavement and drops down to the bed of the river. The river bubbles along here in seeming normality, but

Figure 2.9
The complete Cambrian–Ordovician succession all along the Moine Thrust Zone (modified from Anderton et al., 1979). The Sailmore upwards are only seen in Durness.

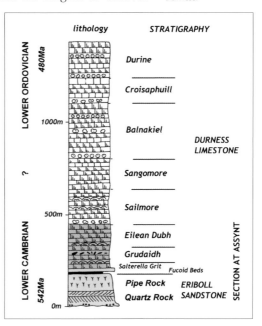

just upstream it has vanished. There is a long subterranean river course through limestone caves well up into the hills, even though on the surface a dry river bed wanders uselessly up the valley too. When the water finally flows back up to the surface it squeezes out between an irregular, nobbly mass of very dark limestone lying uncomfortably on the smooth surface of a bright white limestone. The contact looks wrong, and indeed the water forces itself out precisely along a thrust plane. In geological terms, where the water flows, the dark Ghrudaidh Limestones have been thrust over the white Eilean Dubh Limestones; the natural sequence reversed. In a normal succession, as seen in Stronchrubie cliff, the light limestones form the sheer top cliff and come naturally above the dark limestones. This local thrust plane can be followed right up the dry river bed, the two behaving like inseparable miscreants, until they both disappear into a boggy gully. It is one of the many minor thrust planes that cut through the limestone, here confusing the sequence into a multitude of slices, a thrust fault at the base of each. There is more confusion nearby.

Looking northeast up towards the shepherd's cottage, white rocks glisten from the opposite flank of Gleann Dubh. From their position these should still be limestones, but on close inspection they are the distinctive hard quartzites of the Pipe Rock. The thin streaming waterfall in the Allt Poll an Droighinn, a short way above Glenbain cottage, is a cascade down limestones and 20m or so of Pipe Rock, on top of limestones and some 150m above the loch. In a normal succession of course the Pipe Rock is below the limestones, but hidden in the bog we have just passed over the Glencoul Thrust, which has pushed the Pipe Rock over the limestones, again reversing the normal succession. In fact the entire flank of the hill, up to the face of Conival itself, is still Pipe Rock, and the path up to the mountain-top winds continuously, through sharp boulders and cliffs of the distinctive quartzite, right to the top, to over 975m (3200ft). This is what made the old geologists believe that there were two quartzite layers; one at loch level and another at the top of Conival. Looking up to the triangle of Conival in the distance, the light stripes on the face are all Pipe Rock quartzite with the Glencoul Thrust plane below. A broad, dark, interbedded stripe is an igneous sill within the thrust sheet. This means that in the middle of the Assynt Window, even at 975m, quartzites are outcropping just as they were at loch level by the junction at Skiag Bridge.

The climb to see all this demands an effort, but on a fine day the summits of Conival and Ben More reward with magnificent views: west over Loch Assynt to the waters of the Minch and east to the Moray Firth, both sides of the county in one stunning panorama. The contrast could not be greater with Knockan Crag, which is outside the Assynt Window and a few miles to the south.

Knockan Crag is a National Nature Reserve owned and operated by Scottish Natural Heritage. It is one of the few established predominantly because of its geodiversity interest and importance. The photograph on the front cover is taken from beside the open air exhibition space, the Rock Room, and shows Ben Peach and John Horne, no doubt in deep discussion about the complex history of the Moine Thrust Zone. A trail goes up to the Moine Thrust from here, where mylonitized (ground up and stretched out) Moine Schist has been thrust over Durness Limestone. You can even get under the overhang of the upper thrust block. The fault line can then be traced along the crag above the trail. The path goes all the way along the top of the crag, including a viewpoint, giving wonderful vistas north into Assynt and west to Coigach.

The site had 15,000 visitors in 2018, which is an increase from the average of 10,000 in recent years. It is the most visited site in the North West Highlands UNESCO Global Geopark. Most visitors are independent travellers following the route around the north of Scotland, which has been christened the North Coast 500 in recent years. Those with a sympathetic eye to this landscape already treasured the route before the naming. More visitors undoubtedly give the potential for more learning from this landscape, but could also start to degrade the experience for all. Time will tell whether it matures and is beneficial or not. There are organized groups from

Figure 2.10
The Moine Thrust at Knockan Crag. The dark coloured, metamorphosed Moine schists are thrust over the lighter, Cambrian limestones. The plane itself here is knife sharp. Cliff 40m high, looking east.

cruise boats that call in at Ullapool, and specialist geological tours. Providing information appropriate to such a diverse range of visitors is not easy, as university geology groups obviously have a different need to, and considerably greater prior knowledge than, the casual visitor. The displays around the Knockan trail give the background to why this is such an important site. Don Shelley, the first warden at Knockan, developed the original trail here and amassed a large collection of both local and high quality rock, mineral and fossil specimens from the rest of the world. His collection has recently been secured for a proposed purpose-built centre at Scourie. The Rock Stop centre is currently operating at Unapool near Kylesku and is run by the geopark. It is certainly worth a visit, with an exhibition about the geology and landscape plus coffee, shop and toilets.

Within the geopark there is a series of 14 interpretation panels in lay-bys along the main driving route through the geopark, called the Rock Route. Each takes a particular geological or landscape theme and shows an explanation or interpretation of what the viewers see in front of them. Learning cannot just be about reading and imagining; it must also be about seeing and interpreting for oneself.

Knockan is an ideal locality to consider the scientific controversy that took place in this part of the Highlands at the end of the nineteenth century. Reputations of major geologists of the time were put on the line, as arguments raged about how the structures had formed. The human conflict started taking over from the scientific one, which delayed the understanding of the structure of the area.

Looking west from the centre, the shining rock-covered slope rising to the top of Cul Mor, the mountain opposite, is a continuous surface of the familiar, white, Basal Q uartzite. The rocks are all tilted regularly towards us at about 15°. The path by the centre begins a little higher in the succession, within the Pipe Rock, and on the topmost surface a large piece that perfectly shows the pipes has been prised away and displayed. As the path continues upwards, the brownish, irregularly layered Fucoid Beds make small cliffs, especially in the minor waterfalls to the side of the path, and in one of these the layers have been cleaned and labelled. The fucoids show up well, as does the overlying Salterella Grit, which is usually hard to identify. There is no doubt when the dark grey, rubbly Ghrudaidh Limestones are encountered at the top of a steep zig-zag as the path continues to climb up through younger and younger beds. The limestone is bleached white on weathered surfaces and looks very broken up, fractured and brecciated. The path follows the white, bleached limestones for a while, but half-way up a set of rock steps, they are dramatically and unnaturally cut off like a severed head, by very dark, hard, layered rocks. These are the Moine Schists, crystalline metamorphic rocks, and where they lie in contact with the limestones is the famous Moine Thrust. The thrust plane is perfectly visible for over 90m, knife sharp and simple. Incredibly, on top of this simple surface, all the rocks of the Highland mass have moved at least 80km. No wonder the old geologists were misled. Looking at this surface, even now, it is impossible to imagine the huge mass of the Highlands sliding slowly westwards over it, rock grinding on rock, crystal shattering crystal; but the geology tells us it is so.

In that short 60m climb at Knockan Crag, the Cambrian quartzites, the Cambrian limestones and the Moine Thrust are all passed. Looking up at the top of the cliff from the car park there are the Moine Schists, while such a short distance below is the Pipe Rock. In the Assynt Window a climb of 975m did not even reach the Moines. The geology of the Knockan section is actually more typical of the Moine Thrust along its 160km length. There is the one, major, Moine Thrust that brings the Moine Schists over the unmoved Cambrian quartzite-limestone, Torridonian and Lewisian succession below. In Knockan, the only rocks to have moved are the dark-layered schists in the topmost cliff. Beneath, the succession is quite normal with older below, younger above and all dipping gently to the southeast. At Knockan the thrust has cut right down into a level within the Ghrudaidh Limestones, which explains the very short climb. From the top of the Knockan Crag viewpoint looking the few miles north to the Assynt Window, the peaks of Ben More and Conival tower high above. The climb up Conival showed that within the window the Cambrian rocks have been gathered up into a series of piled-up thrust sheets, and that the normal geological succession is repeated several times. Of course, to the early geologists like Macculloch and Murchison, this was still a natural succession and there was an upper and a lower Quartz Rock, an upper and a lower gneiss, and so on. It was only Nicol and a few other workers who had enough insight to realize that these lithologies might be repetitions of the same thing. We now know that these several thrust slices of rock, or nappes, as they are properly called, formed progressively as movement took place. At first the mass of metamorphosed Highland rocks moved slowly north-westwards, simply sliding on the Moine Thrust, riding on the rocks below as at Knockan. At Assynt, however, as the thrusting continued, the rocks below became snagged and the Cambrian, Torridonian and Lewisian were scooped up into separate thrust-bound sheets. Each

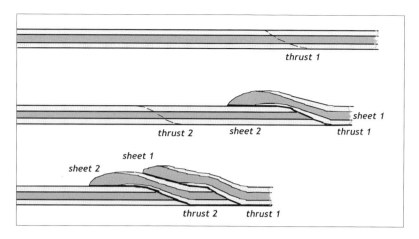

thrust snagged on the back of the one below, the topmost first and each sliding on its own basal fault like the main Moine mass above. The development of thrust sheets at Assynt, one piled on the back of the other, explains why the fault zone is so broad and so high. They indeed create a window, with the thrust-up sediments poking through and framed by the surrounding Moines. Looking at the shape of the Moine Thrust from above, it forms a crescent with a convex outside frame pointing against the direction of movement, that is, to the southeast. This is because the thrust sheets in the window have actually scooped out the rocks underneath, like soft mud pushed up in front of a skidding tyre. Some of these younger, secondary thrusts were identified by Peach and Horne and carefully mapped, being continuous enough to be recognized, such as the Ben More Thrust, the Glencoul Thrust and the lowest of them all, the Sole Thrust. Inside each of the thrust slices there are many minor faults, as in Gleann Dubh, with local bed repetitions mirroring the effects of the larger scale.

When this complex picture is put into context using modern ideas on thrust faulting, it becomes understandable, even if the rocks themselves still remain very complicated. Historically, what caused the early geological errors was that even through all this complexity, there is always a simple, consistent and gentle dip to the southeast. A nice example of the perversity of nature. The modern explanation of the geology of the northwest Highlands satisfies the observations and observers; it is both simple and complicated. Now we can return comfortably to the controversy that we left with an impatient Sir Roderick Murchison. It is still assumed by all geologists that the rocks of the Highland Mountains are Primary and simply framed, on both the east and west fringes, by Secondary red sandstones.

The complication of limestones and quartzites on the west is not resolved.

In 1854 the limestones in Durness provided a huge surprise and the stimulus needed for a long overdue re-examination of Macculloch's work, unchallenged by then for almost 20 years. In that year, Charles Peach (1800–1886), the coast guard at Wick and father of Ben Peach, was inspecting a wreck in Loch Durness but at the same time indulging his passion of looking for fossils. To his delight, by the auld kirk, he came across fossils in the Durness Cambrian limestone, which according to the lithological reasoning of Macculloch was the Mountain Limestone (Carboniferous). He soon sent his finds away to be identified, to Murchison in London, who had just become director of the British Geological Survey. They were poor specimens and difficult to date, but Murchison thought that they might be late Devonian, perhaps older.

What did these fossils imply? Murchison opened the old notebooks he had filled up 30 years before when he had made his only previous visit to the northwest Highlands in the company of Adam Sedgwick. He found that in principle, the fossiliferous limestones were in a normal sequence below the crystalline mass of the Highlands, and were hence older than the rocks of the mountains. There was an immediate conflict in this: the Highland mountains were considered to be Primary, hence pre-dated fossils, and were very old and could not be younger than any fossiliferous beds. Recently though, Murchison had worked on the beautifully fossiliferous Silurian sediments of the Southern Uplands, and he was beginning to speculate about whether or not these rocks continued northwards beyond the Midland Valley. Now, looking at his old notes and perhaps referring to Macculloch's map of Scotland, he came to wonder if the mountains of the Highlands were indeed Primary and very ancient, as had always been thought.

If, on the contrary, these crystalline rocks should prove altered equivalents of Silurian strata, I see nothing but what is rational

wrote Murchison to Sedgwick when discussing the implications of Peach's fossil find. It was enough of a possibility to draw Murchison back to the northwest Highlands, for if the mountains should be Silurian, his empire could be significantly extended. He needed proof.

In 1855, at the age of 63, Murchison always took another geologist to accompany him in the field. It was certainly wise. For that summer he chose James Nicol (1810–1879), aged 45, then a professor at

Aberdeen University. Nicol had worked with Murchison for several years in the Silurian sediments of the Southern Uplands and eleven years earlier, in 1844, had published a *Guide to the Geology of Scotland*, although until then he had seemingly never before visited the northwest Highlands. The object of the trip, of course, was to examine the location of the new fossil finds made by Charles Peach along the north coast. Murchison also intended to look again at the localities in Assynt further south that he had visited 30 years before with Sedgwick – hence the old field notes.

The Highlands were to behave as ambiguously as ever that year, and the weather was so bad when Nicol and Murchison reached Ross-shire and Sutherland, even though it was August, that Murchison got a fit of melancholy. In his diaries he wrote:

> *Old Dunlop dead: my kind old friend Anderson of Rispond, who had sheltered Walter Scott as well as Sedgewick and self, dead. All around gave note that my day was fast coming, and that I had taken my farewell look at Whiten and Far-out Heads.*

This was not the sort of attitude with which to get enthusiastic about rocks, so the trip achieved nothing of importance. The older ideas from 1827 were simply re-instilled, that is, the Highland mass was still regarded as being Primary, although in effect of unknown age, but certainly older than the Old Red Sandstone with Devonian fossils of the east coast. The red sandstones of the west (today's Torridonian) were still regarded as Old Red Sandstone with a similar Devonian age, so nothing had changed. But some detail needs to be added, for even at this stage Murchison had fairly certain beliefs; they progressively became prejudices.

Based on his 1827 visit 30 years previously, and presumably also on Macculloch's work of about the same time, although he implied that it was from recent observation, Murchison was sure that the Quartz Rock and limestones at Eriboll, close to the fossil finds, passed upwards normally into the main crystalline schist mass of the Highlands (Moines), thus dating the schists as younger than Devonian, if this was indeed the age of the limestone fossils. Obviously this was unreasonable because the crystalline sediments on the east were covered by Old Red (Devonian) Sandstones. But Murchison already thought that the limestone fossils should be older, thus leaving open the possibility of a Silurian age for the mountains. The red sandstones of the east, which clearly lay over the schists, were definitely

Devonian, Old Red Sandstones. Moreover, because of his proposed sequence, Murchison was confused as to whether the red sandstones of the west, which he also called Old Red Sandstone (today's Torridonian) were above or below the Quartz Rock and limestone with fossils. Nicol was in no doubt – they were below. The 'facts' appeared to conflict badly. The 1855 visit was clearly unsatisfactory to Murchison and to Nicol, and showed that more work on the outcrops was needed. Murchison had other commitments and was only able to get back to the Highlands two summers later, in 1858. Nicol, on the contrary, did have the time and continued to work in the field.

Without Murchison and the bad weather, Nicol was able to look at the rocks more carefully, and in fact worked his way down the entire outcrop from Loch Eriboll to Skye. He made west to east sections where he could, between Durness and Eriboll on the north coast, in the Assynt–Glencoul area, along Loch Broom and finally near Loch Hourn opposite Skye in the very south. He began to publish this solo work of his from 1856 onwards. First of all he established that the Quartz Rock (Cambrian) is definitely above, and younger than, the red sandstones (Torridonian, Precambrian) and corroborated another geologist's observation (that of Henry James), that the contact is an unconformity, the Quartz Rock having a much steeper dip than the red sandstones (Torridonian), a feature repeated on all sections from the far north down to Skye (just as we saw east of the Inchnadamph Hotel). Nicol also began to look at the contact between what he called the eastern gneiss, that is, the crystalline Moine Schists of the Sutherland mountains, and the underlying limestones. Although his sections at the time generally show a conformable contact, in his text he muses:

> The occurance of igneous rocks [...] at many points along the line of union, seems to indicate a fault.

However, he also says that this seems far too long a feature to be a fault. Nonetheless, he came to suspect that the contact was not a normal one as Murchison supposed, which would mean that the age of what he called the eastern gneiss, the crystalline schists of the Sutherland mountains, could not and should not be judged from the age of the limestones.

Murchison was clearly put out by this, since with additional fossils found by Peach, he had received a fairly good age identification of Lower Silurian (today's Ordovician) for the limestones at Durness.

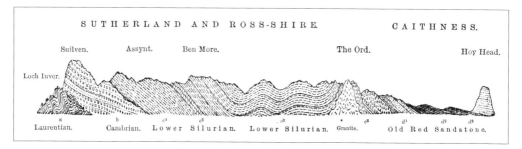

Figure 2.12
A geological section
across the entire
Highlands from the
east coast to the west
based on Murchison's
interpretations.
The Highland mass
of Sutherland and
Ross-shire is entirely
Silurian according
to Murchison! (after
Geikie 1875).

He could then very reasonably propose, as he did in 1858, that the
crystalline schist mass of the Sutherland mountains was altered
Silurian, with Lower Silurian limestones below in a normal succession
to the west, and overlain to the east, across an unconformity, by the
well-dated Devonian Old Red Sandstones. He published a small 'sketch'
map of the whole of northern Scotland to accompany this claim. It
made simple, satisfying Silurian sense. It also meant that not only was
the age of the Sutherland mountains solved – it was Silurian – but
the age of the entire mountain mass of Scotland was equally solved:
Silurian. As Geikie wrote in his biography of Murchison in 1875:

> *And thus in one bold dash of the brush, bold, but justified
> by careful and accurate observation!!* [my exclamations] *he
> wiped out the old conventional mineralogical colouring, which
> dated from the time when gneiss, mica-schist, and clay-slate
> were supposed to be necessarily of higher antiquity than any
> fossiliferous rocks, and substituted for it a mode of representation
> whereby the great mass of the Scottish Highlands was shown to
> consist of altered crystalline sandstones, shales, and other strata
> of Lower Silurian age. No such rapid and extensive change had
> ever before been made in the geological map of the British Isles.*

Never before, never since and, hopefully, never again. The error was
monstrous and Geikie was to suffer for it.

Unconvinced and undeterred by Murchison, Nicol quietly continued
his work and observations. In fact, after the publication of Murchison's
map, he became more and more certain that the amount of faulting in
the northwest Highlands was being seriously underestimated. Initially
he looked carefully at the contact between the limestones and what he
called the eastern gneiss (Moine) all along the exposure. In the Eriboll
area he found an igneous rock between them and actually along the
contact: in the Loch Broom area near Ullapool, something similar
occurred. In fact wherever Murchison showed apparent conformity,

Nicol showed a contact with igneous rocks and a presumed fault. In addition, he looked at the sediments above the red sandstone–quartz rock unconformity (base Cambrian) and proposed that there was in fact a relatively simple sequence: quartz rocks – shale rocks (fucoid beds) – limestones. Any differences from this order, and there were many, were due to faulting. When this was contested by Murchison, Nicol described sections in which beds were overturned. For example, on the west-facing slope up to Ben Arnaboll above the eastern shores of Loch Eriboll, he found inverted Pipe Rock, part of the Quartz Rock interval. The pipes, the fossil 'worm burrows', are here trumpet-shaped with a larger end at the top where the animal fed. In the Arnaboll section, the trumpet ends point downwards. The beds are overturned, clear evidence of important dislocation. It was in this context, coupled with the presence of igneous rocks, that Nicol interpreted the contact between the limestones and what he called the eastern gneiss. He came to regard it as everywhere a great fault, although as we have seen at Knockan Crag, this is not always easy to accept.

Once Murchison had made the 'bold' proposal of a Silurian age for the entire Highlands of Scotland, he became a prisoner of his own ideas. To quote again from Geikie:

> *He stuck to his leading principle, from which no amount*
> *of contradictory detail would make him swerve.*

In truth this refers to another occasion, but was very typical of the man. He was now in battle array, Highlands Controversy array. Not only did he need his social contacts in London to consolidate the acceptance of his interpretation but, because of Nicol's persistence, he

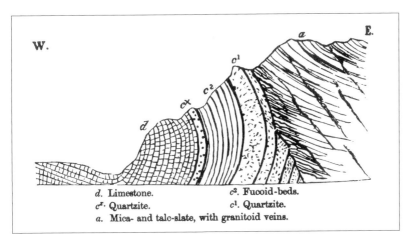

Figure 2.13
An early section from Loch Eriboll near Durness on the north coast by Nicol. He found inverted Pipe Rock here (after Nicol 1861).

also needed geological evidence from the rocks themselves. So, in the summer of 1859 Murchison went back to the Highlands, this time with Andrew Ramsay (1814–1891), his second in command at the Geological Survey, and they stayed a month or so between Ullapool and Eriboll with visits to Assynt and Loch Glencoul. Most of the time they spent checking up on Nicol's proposals, checking to see if there was indeed igneous rock between what he referred to as the eastern gneiss (Moine Schist) and the limestones. Murchison managed to convince Ramsay even while in the field (forced Ramsay to agree is probably not fair), that there was no such thing and that the succession from limestone to eastern gneiss was everywhere natural and conformable.

During the 1800s, new scientific advances were presented every year at a meeting of the British Association for the Advancement of Science, held in different towns around the country. All the sciences were involved but the natural science and geology section was especially important at the time. In 1859 the Association meeting was held in Aberdeen and the geological section was organized by Nicol as he was the Professor of Geology there. Murchison and Ramsay travelled to Aberdeen immediately following their work on the outcrops of Sutherland and Ross-shire. Hot from the rocks, Murchison presented his interpretation of the northwest Highlands, which, as expected, was the simplistic one that he had always proposed, of a normal easterly-dipping succession, from Quartz Rock and limestone into eastern gneiss (Moine Schists) with some simple repetitions of the various lithologies. For all to judge, the Highlands were Silurian, albeit by inference. Such a clear and simple explanation of the geology was understandable by everyone who listened to it, especially those, the great majority, who had not visited the area for themselves. And so it was accepted, by opinion poll as it were, that this was the correct explanation, especially as it was proposed by the substantial Murchison and seconded by the also-reputable Ramsay. The Aberdeen Journal reported:

> So plain and irresistible were the evidences in a number of
> transverse sections at Loch Broom, Loch Assynt, Loch More,
> Loch Eriboll, etc., that the only wonder entertained was that any
> scepticism should have prevailed as to the order of succession.

Murchison had the foundations of both a social and scientific victory. Poor Nicol, busy administering the meeting, presented his interpretations as well, but was only able to do so much later in the sessions. By this time Murchison's success held sway and his interpretations had been entirely accepted so that coming second, with lesser social and scientific standing and especially with a far more complex theory, he did not stand a chance. In 1859 Nicol lost to the immediate audience but his own scientific integrity remained intact. Regrettably however, calumny was already on call. When the proceedings of the Aberdeen meeting were published, Murchison's paper was nowhere to be seen, Nicol's was printed in full. As David R Oldroyd, who has written an excellent book on these events, comments:

> *one is naturally inclined to suspect [...] Nicol.*

Despite his difficulties at the 1859 meeting, Nicol continued to do original work and prepared his further observations for publication, his next paper being read to the Geological Society in December of 1860. In it, he reinforced his previous observations and proposed a sequence upwards of: gneiss (today's Lewisian), unconformity, red sandstone (Torridonian), unconformity, Quartz Rock, some with pipes, fucoid beds and limestone (Cambrian) fault with igneous rock and then eastern gneiss or schist (today's Moine), in fact exactly the sequence accepted today. Any difference in this order was caused by tectonics, and as evidence of this the overturned Pipe Rock on the east side of Loch Eriboll was cited. Nicol also detailed his observations on the upper contact, explaining the presence of schist or gneiss above the limestone as the result of faulting, the fault being picked out by some sort of igneous rock.

Although Nicol's basic theory was, remarkably, correct, he also made a number of errors. The faulting that he envisaged was high-angle and near vertical; horizontal thrust faulting had not yet been observed or proposed and was a minor consideration at that stage. He made a more serious error, however, when he indicated the lower, western gneiss (Lewisian) and what he called the eastern gneiss (Moine Schist) to be similar, despite the fact that he had not taken this view originally and that the 'tectonic grain' of the two rocks is quite different. In his defence, he writes in one article that this similarity is in being metamorphic, age not being implied. An additional minor point is that some of his so-called 'igneous' rocks are not igneous but a mixture of types including mylonite, the rock mechanically altered by thrusting, although at the time this type of rock was unknown. The undoubtedly important proposal that he made, of course, was that nowhere was there a normal succession of rocks, as Murchison strongly maintained, between the crystalline Highland mass and the dated limestones. A Silurian age for the huge area of the Highlands could not be scientifically implied, if his (Nicol's) hypothesis was correct. Moreover, he pointed out quite correctly, that even if the eastern gneiss did overlie the limestones in a normal succession, this was absolutely no basis for saying that the whole of the Highlands themselves followed this simple succession; they were far too huge and complicated. As for Murchison's sweeping proposal to put a Silurian instead of a Primary age across the Highlands he retorted 'No such revolution in Scottish geology ... is required.' But he was only a distant Scottish voice in a ruling English establishment.

Nicol's paper containing all this left Murchison more than displeased. Normally, a paper read before the Geological Society would be immediately published in the Society's transactions, but Nicol's paper did not appear in print for another 14 months, until February 1861, a delay quite satisfactory to Murchison. He used (caused) this delay to consolidate his own field work and publish his own ideas before those of Nicol. Revenge for the omission in the reporting of the Aberdeen meeting?

The field trip Murchison made in 1860 when he was 69 was to be his last anywhere. Although each time he went to the northwest Highlands Murchison's partner and his itinerary were different, his objective was always the same; to prove that the Highlands were Silurian and to discredit any work that Nicol, amazingly still his only contestant, may have done to counter his proposal. For this last trip Murchison took with him the young Scottish geologist Archibald Geikie (1835–1924). Their time together in the field founded a very strong friendship, Murchison seeming to treat the young man as the son he never had. At the time of the trip, Geikie was an extremely ambitious 25-year-old working for the Scottish section of the Geological Survey and although the friendship on both sides was genuine enough, Geikie used Murchison to further his career while Murchison, who had complete domination over Geikie, used his youthful enthusiasm to further his Silurian cause. Geological observation did come into this collaboration somewhere, but seems to have had only a secondary influence on the outcome. The chapter in Geikie's biography of Murchison that covers this period, for instance, is entitled 'The Geological Conquest of the Highlands'. It was, of course, nothing of the sort. A geological section of the Northern Highlands published in the third edition of Murchison's *Siluria* in 1867 shows the Lower Silurian age of the main mountain mass and the overall geological structure clearly; it is clearly wrong (cf. Figure 2.12). The observations and interpretations of Nicol had been ignored for the sake of Murchison's Silurian Empire.

Today's physical wars are limited by money. The Israeli Six-Day War against the Arabs in 1967 was limited by financial resources. The Falklands war cost a huge amount: the war in Yugoslavia is still being paid for by the West: the Iraq war they will pay for themselves. Academic war has no cost; it is limited only by appetite. From 1860 on, the appetite was clearly Murchison's and Nicol, after his original struggles, made little effort to engage. It is clear that the war was political and not scientific. Murchison had excellent political ordnance, Nicol had only good scientific ordnance. With his science Nicol might have been rescued but, as Geikie wrote in his 1875 biography of Murchison:

> *With this Highland tour (1860), and the preparation of the last narrative of it for the memoir (in 1861), Murchison closed the last great geological task of his life.*

Murchison took his errors to the grave; he died in 1871. Silurian colours were to stain the geological map of Scotland for the next 20 years with no one daring to wipe them off.

After his death, Murchison's interpretation of the Highlands became Geikie's. He carried on the Silurian cause, first as both head of the Scottish Geological Survey (1867) and Professor of Geology at Edinburgh University (1871) and then eventually as head of the entire British Geological Survey (1882) in London. With Geikie in charge of 'Government geology', the official geological interpretation of the Highlands was safe. There was no reason to do more work; there was nothing new to learn. The subject was now no longer debated and the Highlands Controversy was all but forgotten. But it was not extinct: only temporarily dormant.

Towards the end of the 1870s independent, often 'amateur' geologists began to look at the northwest Highlands for the first time and discovered, to their surprise, that the official Geological Survey interpretations, that is, Murchison's, were barely tenable, almost certainly wrong. Papers began to appear questioning the official Survey view being upheld by Geikie. The first in 1878 was by Henry Hicks (1837–1899), but this was quickly followed in 1880–81 by a paper from Charles Calloway (1838–1915), who used detailed geological field observations to show that the Government view was wrong. Nicol, although he met the dissenters, was now old and did not get involved directly. He died in 1879. But with a new flow of papers the Highlands Controversy erupted again, although this time going in the deceased Nicol's favour. Calloway's challenge to the official, Murchison view was countered by the Government geologists, directed by Geikie, who took the high-handed attitude that the Government knew best and that the amateurs were not to be believed. This was the riposte every time there was a challenge.

This time though, the contest was to be different. The ghost of Murchison lived on in Geikie as the Director-General of the Geological Survey, but instead of the lone voice of Nicol on the other side, as there had been before, there were now ranged a vociferous number of professors, independent academics and the amateurs, all with an appetite for contest. The politics was now not Silurian, but part of a generally increasing dissatisfaction with the scientific incompetence and arrogance of the Survey, who considered themselves as the country's official voice on geological science. The dissatisfaction focused especially on Geikie himself, the epitome of a tyrannical and isolated manager.

Despite his robust defences, Geikie realized the seriousness of the challenge. Moreover, between 1882 and 1883, Charles Lapworth (1842–1915), at the time professor of geology in Birmingham, made clear and detailed maps of the Durness area. These not only disproved the Murchison version of Highlands geology, but effectively provided the foundations for a solution. Lapworth was familiar with the new theories of horizontal faulting being proposed for the Alps by Swiss geologists and also being researched in America and Scandinavia. He realized that what he was seeing in Durness was the evidence of the westerly movement of the whole Highland mass over the rocks below. The emotional excitement of this discovery was too much for

Figure 2.14
The complex geological section of the Assynt Window produced by Peach and Horne for the 1914 Memoir.

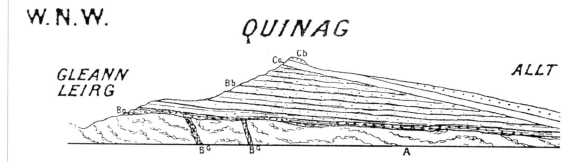

him and provoked visions and nightmares about the whole weight of the Highland mountains grinding slowly over his own body, and he had a severe mental breakdown. But his challenge was enough, and Geikie was forced to start a massive Government re-interpretation of the Highlands led by Ben Peach and John Horne. Their work started in 1883 and continued for 14 years, being completed only in 1897 and culminating in the beautiful memoir of 1907.

Although the final solution to the Highlands Controversy was provided by the Government Survey, it was the non-survey geologists who had seen the original Murchison errors and had first provided new scientific insight into the whole affair. This put the Survey in a very bad light: their competence, efficiency and certainly their authority were seriously questioned. Remarkably, taxpayers became involved, eventually forcing the acrimonious retirement of Geikie himself in 1901. Murchison's ghost had brought him down; he could not separate himself from his mentor and was eventually destroyed by him. It was only with Geikie ejected that Murchison's ghost could be finally pierced through its Silurian heart, and the geological map of the Highlands given its true colours.

It is human nature to search in this story for what is called in America a 'take-home lesson', although to me this sounds rather like philosophical fast food. The moral to be learnt depends on one's leanings. Do we condemn the brilliant but overbearing and socially arrogant Murchison and Geikie? Do we praise the hardworking, self-effacing Peach and Horne or the confident intelligence and originality of Nicol, Calloway and Lapworth? Or do we focus on the way that scientific truth will always triumph? Perhaps none of these, important though they may be. The lesson is: be open to progress. An unnecessary, often desperate attachment to the past is an error. A lesson for science. A lesson for life?

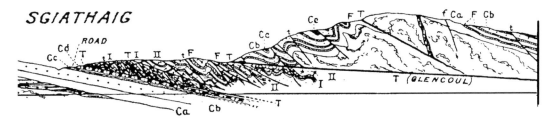

Further Reading

Books, Pamphlets

Anderton, R., Bridges, P.H., Leeder, M.R. and Sellwood, B.W. 1979.
A Dynamic Stratigraphy of the British Isles. London: George Allen & Unwin.
pp.301. ISBN 0 04 551028 8

Bailey, Edward. B. 1952. *Geological Survey of Great Britain.* London: Murby.

Cadell, Henry M. 1896. *The Geology and Scenery of Sutherland.* Edinburgh:
M'Farlane & Erskine. 2nd Edition, pp108.

Dryburgh, P.M., Ross, S.M. and Thompson, C.L. 2014. *Assynt: the geologists'
Mecca.* Edinburgh: Edinburgh Geological Society. 2nd edition, pp 33.
Also available from the EGS website as a free download.

Geikie, Archibald.1875. *Life of Sir Roderick Impey Murchison.* 2 Vols.
London: John Murray.

Goodenough, K., Pickett, E., Krabbendam, M. and Bradwell, T. 2004.
Exploring the Landscape of Assynt. British Geological Survey Publication.
pp.55 (with map). ISBN 085272471 3

Johnson, M.R.W. and Parsons, I. 1979. *Geological Excursion Guide to the
Assynt District of Sutherland.* Edinburgh Geological Society. pp.76.

Johnstone, G.S and Mykura, W. 1989. *The Northern Highlands of Scotland.
British Regional Geology.* 4th Edition, pp. 229. British Geological Survey,
Her Majesty's Stationery Office. pp.219.

Krabbendam, M., Strachan, R. (eds). 2010. *Continental tectonics and
mountain building – the legacy of Peach and Horne.* London: Geological
Society, Special Publication No. 335. pp. 880.

Macculloch, J. 1819. *A description of the western islands of Scotland,
including the Isle of Man, comprising an account of their geological
structure, with remarks on their agriculture, scenery, and antiques.*
3 Vols. London.

Oldroyd, David R. 1990. *The Highlands Controversy.* Chicago: The University of Chicago Press, pp.438. ISBN 0 226 62635 0

Peach, B. N., Horne, J., Gunn, W., Clough, C.T., Hinxman, L.W. and Teall, J.J.H. 1907. *The Geological Structure of the North-West Highlands of Scotland.* Memoir of the Geological Survey of Great Britain, HMSO. Glasgow. pp.668.

Phemister, J. 1960. *Scotland: The Northern Highlands. British Regional Geology.* Geological Survey and Museum. Her Majesty's Stationery Office, Edinburgh. pp.104.

Scientific Papers

Butler, R. W. H. 2004. The Nature of 'roof thrusts' in the Moine Thrust Belt, NW Scotland; implications for the structural evolution of thrust belts. *The Geological Society of London* 161, pp. 849–860.

Butler, R.W.H. 2007. Peach and Horne – the memoir at 100. *Geoscientist* 17, pp. 20–25.

Nicol, J. 1861. On the structure of the north-western Highlands and the relationship of the gneiss, red sandstone, and quartzite of Sutherland and Ross-shire. *Quarterly Journal of the Geol. Soc. London.* No.17, p.85–113.

Chapter 3   THE FISH GRAVES OF ACHANARRAS

Devonian fish are human ancestors

My advice to young working men desirous of bettering their circumstances, and adding to the amount of their enjoyment, is a very simple one. Do not seek happiness in what is misnamed pleasure; seek it rather in what is termed study. Keep your consciences clear, your curiosity fresh, and embrace every opportunity of cultivating your minds. [...] Learn to make a right use of your eyes: the commonest things are worth looking at, – even stones and weeds...

There is no other way to start this chapter than in Hugh Miller's own words, although it is hard to imagine them today as they were then, an introduction to his book on geology. But this book, *The Old Red Sandstone*, had tremendous success. It was first published in 1841 and was still being reprinted at the turn of the century 60 years later, in its 26th edition. This chapter is not about Hugh Miller, though,

Figure 3.1
A portrait of Hugh Miller. He started as a simple quarryman but became a renowned expert on the Old Red Sandstone of Scotland (Engraving by Bell of a photograph by Tunny, courtesy of the Hunterian Museum and Art Gallery).

but about the fossil fish of the Old Red Sandstone of Caithness, and incidentally about stratigraphy and especially evolution. Hugh Miller is in this chapter because he spent 20 years breaking the rocks of the Old Red Sandstone as a quarryman; nobody has done that since or is ever likely to do it again. He is also here for his lovely prose, which will be used to convey his contagious enjoyment and enthusiasm for science and for life. But first let us briefly introduce him.

Hugh Miller (1802–1856) was born in the Black Isle fishing village of Cromarty, sheltering under the Sutors at the entrance to the Cromarty Firth, 60km north of Inverness. He became a quarryman at 15 and then, by sheer intelligence and force of character, some 20 years later had become the authority on the geology of the Old Red Sandstone in Scotland, especially the Highlands. He became fully embraced by the 'socially superior' geological community in London, a remarkable achievement in Victorian times. He was a very religious man, an excellent writer, and became a renowned Edinburgh newspaper editor. His geological classic is *The Old Red Sandstone* mentioned above, but he also produced a number of other geological works such as *Footprints of the Creator* (1849) and *The Testimony of the Rocks* (1857), all being an inextricable mixture of rocks and religion, of nature living out God's will. But this is not to underestimate his science: he was an observant, gifted and enthusiastic scientist. Miller was not only unusual in his background, he was physically so as well; tall, rugged, with a full head of shaggy sandy hair, and usually dressed roughly in a plain shepherd's plaid, preserving the image of his origins. 'He never looked grander than a working man in his Sunday best,' said a friend. He was a distinctive personality and a great speaker with a broad Ross-shire accent, keeping audiences of thousands - especially ladies, it seems - rapt with descriptions of rocks and fossils.. He was a thoroughly likeable character and when he died in Edinburgh, the centre of the city was closed for his funeral. From now on he can use his own, distinctive words.

The place where we are going to start this chapter has been designated a Site of Special Scientific Interest (SSSI), and to visit needs a permit from Scottish Natural Heritage (SNH). It is an inconspicuous quarry near Spittal, 24km south of Thurso, at the end of a long farm track on top of a low hill in the searchingly flat Caithness countryside. The quarry itself is entirely unremarkable at first sight: water-logged, no remaining worked face, a ruined building and a lot of broken rock spoil clambered over by cows. Yet over the years this quarry has provided thousands of beautifully complete fossil fish, it having been excavated unintentionally into the Achanarras fish bed. This is why permission is needed from SNH to visit, hit the rock and to collect the fossils. A few years ago a Dutch team brought power drills and heavy equipment to improve their collecting efficiency! Not allowed; the rules are one hammer and a few fossils.

There is a mystery here, though. The rock is perfectly layered to resemble pages in a book and you can sit in the quarry for hours splitting layer after layer and find absolutely nothing. The excitement that each split may reveal beautiful shiny black scales and the first look at a curious ancient fish is rapidly lost. But with experience and some knowledge of collected specimens, the fossil-hunter learns to look for black or rusty red weathering colours in the grey-green rock. The truth is that the magnificent fossils really are abundant, but only in a very precise layer two metres thick, hence the name 'fish bed'; the fossils are not scattered throughout the rock. Fish beds are now a known feature of the Old Red Sandstone of Caithness, Cromarty and the southern shore of the Moray Firth, but it originally took Hugh Miller years to find some of them even though he was on the rocks every one of his working days.

Why are the fossils so distributed? Why are they concentrated only in the thin fish beds? These are the kinds of question that give the background to this chapter. Do the thin layers of abundance represent a time when a huge number of fish lived? Or do they represent the catastrophic death of a whole fish population? Perhaps there is something special about their fossilization?

The different degrees of entireness in which the geologist finds his organic remains depend much less on their age than on the nature of the rock in which they occur

was Hugh Miller's surprising (and accurate) observation. What was special for him, evidently, was the fossilization. But we should try to find out for ourselves, because it affects our understanding not just of these fish fossils but of the entire fossil record, what we say about evolution and what we give as the age of a rock. Fossils cannot be considered in isolation; they are inextricably linked to the sediments in which they are found, 'the nature of the rock'. The sediments represent both the environment of the living animals and the medium in which they are preserved. However, before trying to make sense of the Achanarras fossil fish, I shall briefly discuss stratigraphy, already introduced with the red Torridonian in Chapter One.

Stratigraphy can be lightly defined as the study of geological past history, having as its main objective putting sedimentary rocks in order of age. This may be the objective today, but it was not the same 200 years ago when the study of 'historical geology' first started, as stratigraphy was then called. Examining the past frequently helps to explain present-day illogicalities. Why do men wear ties, for instance? They were first worn by Croatian mercenaries in the seventeenth century, quickly became a fashion item and remain so even today. They were clearly not initially the sign of conventional respectability that they are now. A similar 'ghost from the past' is very much present in the practice of stratigraphy.

The objective of early stratigraphic work was to define the Universal Sequence that showed how the Earth had been created. Primary, Secondary and Tertiary are the stratigraphic ghosts from this time, and terms which are still in use today, but in meaning much modified. Originally, Primary rocks were identified as being crystalline, mainly massive, devoid of fossils and generally forming high, mountainous ground or mountain cores. Secondary rocks were layered sediments, tilted or slightly folded, containing fossils not recognized today and forming foothills. Tertiary rocks formed low ground, were unconsolidated sediments, were unfolded and had familiar fossils. This simple system had a certain attractive truth about it, and we have seen it used in the previous chapters. The entire Earth consisted of the three, universal layers which eventually became to be considered as a sequence of superposition, one layer over the other, one younger than the other, the Tertiary layer last and youngest (Quaternary was later added).

In the early 1800s when coal was being actively prospected, this simple system was put to the test and found to be rather inadequate. The definition of the Secondary layer especially, where the rocks with coal occurred, became much modified based on practical prospecting experience and was pragmatically sub-divided. At this time it was the 'look' of a rock which counted: in essence its lithology. For example, in the search for coal, the German miner Abraham G. Werner

Figure 3.2
A stratigraphic section of sedimentary rock layers illustrating the principle of superposition. Layer 1 was the first to be deposited, then 2 and so on (after Geikie, 1886).

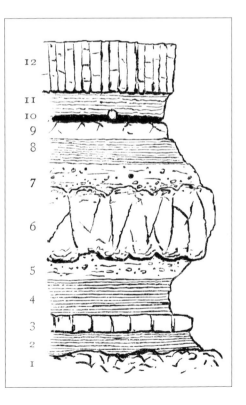

(every student knows that the G is for Gottlob) (1749–1817) was able to suggest that coal-bearing beds would always be found between a red sandstone above, his *Rothe todte liegende* (red lifeless layers) and a limestone below (the Mountain Limestone). In Scotland this was quickly discovered not to be true, and although there was indeed a red sandstone above the coal beds, there was a red sandstone below as well. The one above then became the New Red Sandstone and the one below the Old Red Sandstone. This was one of a number of modifications, and it was becoming evident to the geologists of the day that lithology was not a reliable character for correlation, for recognizing the same layer from county to county, country to country.

The idea of using fossils instead came from the English surveyor and canal-builder William Smith (1769–1839). In the Geological Society of London he is called the 'Father of English Geology', and his famous geological map hangs, rather faded, near the top of a flight of carpeted stairs in the Society's prestigious rooms in Piccadilly, although during his life he was considered by some members to be socially inferior, and not the 'right sort of person' to be a member. The Americans call him 'Strata Smith'. Take your choice how you address him. During the very early 1800s, while building canals around Bath and in the English Midlands, he established the principle of what we now call biostratigraphy. What Smith proposed was recognizing rocks from their fossils, rather than just their lithology. As an illustration, limestones were recognized in many parts of Britain – for example in Derbyshire and Oxfordshire – and before Smith were considered similar simply because they were limestones. What Smith saw was that the fossils in the Derbyshire limestone were quite different from the fossils in the Oxfordshire limestone. The former, what was called the 'Mountain Limestone', has many brachiopods, crinoids and rare goniatites, while the Oxfordshire sediments have abundant ammonites, corals and molluscs. The Mountain Limestone is, of course, Carboniferous while the limestone in Oxfordshire is Jurassic, some 150 million years younger, but for Smith it was the difference that he was able to demonstrate. Subsequently geologists realized that fossils form

Figure 3.3
A stratigraphic section made by Hugh Miller following William Smith's system and with the Old Red Sandstone sequences identified by their fish fossils (after Miller, 1873).

OLD RED SANDSTONE OF SCOTLAND.
SECTION I.

Upper Formation.   Middle Formation.   Lower Formation.

Carboniferous Holoptychius. N.   Cephalaspis. L.   Dipterus. M.   Granitic Gneiss.

an ever-changing sequence, which could therefore be used to indicate the age of strata. 'God never repeats himself' said Miller. Darwin explained this progression in terms of evolution. Darwin, however, did not publish his theory until 1859, 20 years after Smith's death. Nevertheless, the use of fossils quickly became routine for allocating an age to sediments.

Today, using the principles of biostratigraphy, that is simply using fossils, a well-established stratigraphic chart exists that covers the last 550 million years of Earth history, the period over which animals have left hard body fossils. The modern stratigraphic column, as the chart is called, is a list of divisions of all the ages, a kind of modern, composite 'Universal Sequence'. On this, the divisions shown, based on characteristic fossils, are found throughout the world from China to South America and Africa to Europe, although never all in one place. A sediment layer can be examined in China and, using the fossils it contains, be identified as having a certain age and

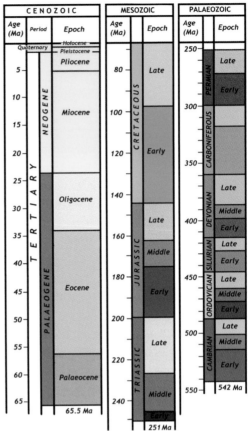

GEOLOGIC TIME SCALE

a certain position in the column, even though it will have a local character and local name. The stratigraphic column acts as a universal reference for rocks, just as the meridian at Greenwich does for time. A few years ago, sediments from a remote area in the Himalayas were said to contain fossils which were totally unexpected. This was astonishing, as such fossils had never been found in this area before. The geologist who found them, like a drugged-up sportsman who suddenly performs spectacularly, became instantly famous. He then became even more famous, like a drugged-up sportsman exposed, when it was found that he had simply bought the fossils in the local bazaar! The mountains were just too far away and too high to actually go and collect them.

This particular geologist is now discredited and the sediments have resumed their previous status, but the importance of fossil evidence is nicely shown.

Figure 3.4
The Stratigraphic Column. The time chart used by all geologists only shows detail since the Cambrian (ages from Gradstein et al., 2004).

In modern geology, we now go beyond fossils and use absolute age dates, as for the Lewisian in Chapter 1 (returned to in Chapter 6). These ages are found by analysing certain radioactive element ratios in crystalline rocks. Consider, however, that in these rocks the atomic clock is set when the mineral crystallizes at high temperatures, usually deep within the Earth's crust. This is fine for dating the cooling of igneous intrusions or metamorphic rocks like the Lewisian, but the technique cannot be used to date sediments on the Earth's cold surface. For this we must use a few tricks. A rock from a volcanic eruption comes straight from molten material deep in the Earth and its atomic clock is set just before it escapes to the surface. This material may be ejected as a lava flow or as volcanic dust in an explosion – the sort that goes into the stratosphere and gives red sunsets for several months. It is this volcanic dust that is best for dating sediments, for as the sunsets testify, it can be scattered over a wide area, especially over the sea, falling to the sea floor and forming a thin sedimentary layer called a tuff or tephra. This technique of dating volcanic ash layers has been used in the Olduvai Gorge and Koobi Fora beside Lake Turkana in East Africa, to give a very precise date to the very early human remains found in these areas. Finding sufficiently thick, preserved, rather rare, volcanic tuff layers through the entire range of geologic strata and then dating them has taken quite a long time. Even now, some of the absolute ages are still $\pm$ a few million years, especially the older ones. But for the most part, there is reasonable agreement amongst geologists so that the stratigraphic column with absolute age limits is now standard geological information. For example, the Old Red Sandstone of our present interest is Devonian and spans 408Ma to 360Ma. Nevertheless, even with absolute dates, fossils still provide the everyday tool for recognizing a sediment's age, be they large fossils like fish, small fossil shells, or fossils only identifiable using a microscope. This is why it is important to understand the fossilization of the fish of the Achanarras quarry; they are the coded instructions for using this tool.

Miller would not have known the Achanarras quarry; it was opened in 1870, twelve years after his death. This is a shame. He would certainly have revelled in the abundance of the fossil fish, although, at the same time, have been surprised and fascinated by the way in which the sediments and the fossils themselves have now been analysed.

What is immediately remarkable about the rocks of the quarry and indeed all the rocks of the Caithness area, is their extreme, regular layering. These are the Caithness Flags, famous for covering floors around Europe, all the way from many Scottish kitchens, to the Royal Mile, to many streets in Edinburgh and now the infamous new Scottish Parliament building. The flags are so readily available in the north that lines of them are used to separate fields, instead of dykes (stone walls) as in Sutherland or hedges in lusher counties. The amazingly flat beds were caused by the exceedingly regular accumulation of sediments in the huge freshwater Orcadian lake that stretched from Cromarty to Caithness to the Orkney Islands, and to the west coast of Norway, during the Middle Devonian, 374–387Ma. With a size of about 700 by 300 kilometres, it was larger than the individual North American Great Lakes and comparable with the size of the present Black Sea. But undoubtedly, during its 13 million years or so of existence, lake outlines shifted and water stands were often temporary, similar to the lakes of the present day African Rift. Over 5000m of stratified flags accumulated during this time, in the general area of the slowly sinking but usually isolated Orcadian Basin. The powerful, vertical cliffs of Caithness and Orkney are the remains of this remarkable sequence, the Old Man of Hoy their famous statue. Incredibly, throughout this whole Orcadian lake sequence the flaggy layering persists. Each flag is usually about the thickness of a paperback book and exceedingly flat, which gives it its economic worth. Although persistent, the layering shows subtle changes related loosely to the silt, sand and shale content. All the fish in the Achanarras Quarry come from a thickness, as I have said, of only two metres, and through this interval the individual layers are extremely thin. The fish beds are in fact laminites, that is, sediments with paper-thin layering. It is only in the laminites that the fish are found: or is it preserved; or is it lived; or is it died?

Although not visible without draining, researchers have found that the sedimentary layers of the Achanarras quarry form a sequence of gradually increasing particle size and lamina thickness from the bottom upwards. Ten metres are found at the quarry, but when complete, it is about 60m thick. The finest particle size and thinnest layers at the bottom of this sequence are the two metres of laminites where the fish are found. They occur nearly at the base of the quarry. The laminites were deposited in still water, deep for a lake, perhaps 80m. The thicker layers at the top of the sequence show fossilized ripples and mud cracks and clearly come from very shallow water, which even occasionally dried out. The rocks in the quarry are the remaining evidence of an episode when the lake became progressively shallower as time went by. It slowly filled with sediment and the lakeshore eventually passed over the quarry site, whereas before it had been a long way off. When the lakeshore environments were present, the water was well oxygenated and any organic remains in them would have been consumed by bacteria or predators or dispersed by water currents. On the contrary, when the deep water was present, the lake floor would have been anoxic, that is without oxygen. There would have been no bacteria or predators and no water movement, which would have been ideal for the preservation of organic matter. This, of course, is where the fish remains are. But did the fish only live when and where the water was deepest? Unlikely: but let's look at the fish fossils and the laminites themselves to find out more.

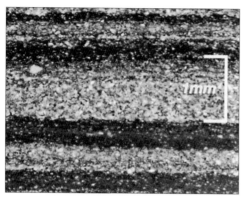

Figure 3.5
The Fish Bed laminites from Achanarras. The white layers are silt and micro-dolomites (CaMg carbonate), black layers are fine, organic rich mud (courtesy of Nigel Trewin).

Carefully polished and acid-etched surfaces of the laminites show that the middle 70cm, the most fossil-rich part of the fish bed, is made up of alternating layers of light-coloured carbonate and dark organic matter of perhaps 0.07mm thickness (that of stiff card). The carbonate, actually calcium and magnesium carbonate, is formed of very fine, microscopic mineral particles. The organic matter is made of algal remains, algae being the microscopic plants that grow on water, making summer ponds green. The climate of the lake was tropical to sub-tropical and each pair of layers is the record of a season, like tree rings. First, probably in midsummer, would come the yearly algal bloom, when a huge number of algae would form a scum over the lake surface, changing the chemical composition of the water and causing precipitation of a fine carbonate 'rain', which would fall to the lake floor to form the carbonate part of the pair. Eventually, normally in the late autumn, the algae would die *en masse*, dropping on death to the lake floor to form the thin organic layer, being preserved because of the lack of oxygen and bacteria on the lake bottom. Added sometimes to these pairs would be a layer of silty material, probably brought in by storms as wind dust or sediment from the lakeshores. Below and especially above the middle 70cm of the fish bed, silty layers are more frequent and fish remains rare; the greater the amount of silt, the less the chance of finding fish. In brief, the laminites represent a time and place in the lake with very quiet bottom waters where fine chemical and organic matter accumulated out of suspension in annual couplets called varves. This description could fit many of the world's larger, hot climate lakes today, but the rock diary we are looking at is over 375 million years old.

And what about the fish? When we look at those that are preserved in the laminites, first of all we find that from any single species there is a mixture of sizes. Then we find that there is a mixture of fish from very different lifestyles. These are noteworthy facts: they indicate mass mortality, the catastrophic death of a whole community, not a normal dying. Miller also recognized this:

> *It presents us, too, with a wonderful record of violent death*
> *falling at once, not on a few individuals, but on whole tribes.*

A normal death community is like a graveyard where there is a preponderance of old people and if the dead were all dug up and analysed, they would not be representative of the living population. Moreover, as we know, different religions have different graveyards; there is no mixing of souls. This contrasts with the dead from a village annihilated by an earthquake, where all die together regardless of age and status; this is the population of the fish beds. Old and young, small and large are found together, as are bottom feeders that lived in shallow water along with free-swimming carnivores unrestrained by water depth.

> *The convulsions and revolutions of the geological world, like those*
> *of the political, – are sad confounders of place and station, and*
> *bring into close fellowship the high and the low*

regrets Hugh Miller, who 'knew his place'. Catastrophic death and the 'confounding of place and station' seem to have been a regular feature in the Orcadian lake. Something was regularly able to kill off large numbers of fish, despite the lake being normally able to contain a thriving living population, as the fossils themselves testify. Moreover, this 'something' seems to have happened at a particular time of the year, a particular part of the light and dark paired sediment layers.

The clue to this 'something' is in the fish beds themselves, and that is that they are just that, only fish. Miller noted this and remarks,

> *In the ichthyolite beds of Cromarty and Ross, Banff, Perth, Forfar,*
> *Fife, and Berwickshire, not a single shell has yet been found.*

(Ichthyolite is the name for fossil fish). The explanation is in the style of the catastrophic death. In some Highland salmon rivers, occasionally and very regrettably, poachers collect their fish by putting strychnine into the water. This has the effect of taking away all the dissolved oxygen and literally drowning them. The fish then float to the surface and collect as flotsam in quiet pools, to be lifted out and sold on to dishonest dealers. In the early morning you can still find fish floating around, untouched from the night before. There were of course no Orcadian poachers, but the slime of an algal bloom has the same effect as strychnine; it starves the fish of oxygen by clogging their gills. In the shallow lakeshore waters where the fish lived, a bloom would have had a devastating effect on bottom-living and free-swimming forms alike. On death, the fish would all have floated to the surface to be caught in currents or blown by the wind to the deeper parts of the lake, where they eventually sank. It was here that they fell to the anoxic part of the lake floor to be fossilized for us to find hundreds of millions of years later. Clearly, only remains that drift would be fossilized, and while other animals that lived with the fish in the shallow waters, such as freshwater shrimps and snails, may also have died with them, since they did not float they do not occur as associated fossils. Hugh Miller was right: conservation counts. Without those catastrophic events, the fish would not have died and we would never have known of the rich life the lake contained in its waters and around its edges.

This sounds like a fisherman's story, a palaeontologist exaggerating about how remarkable his Middle Devonian fossil fish are, how their fossilization is one tiny chance in many billion. It is remarkable. Any fossil is remarkable and any permanent preservation of a living form is unlikely, more so with soft-bodied, land-living or flying forms than with forms living in water. Our own species is an example. The earth groans with humans and yet the number of bodies preserved, especially with their skin intact, is incredibly small. The 2000-year-old people found in ancient bogs and marshes in northern Europe, so amazingly and completely pickled by the acid soils with clothes and perfectly wrinkled skin, are as famous as they are rare. Mammoths from the Siberian permafrost, deep-frozen for 20,000 years with woolly coat and internal organs still intact, even more so. So rare is fossilization that the fossil record might be looked on as a record of freaks – not so. The fossilized forms are quite normal. It is, as Miller said, a question of conservation; it is the conditions under which fossilization occurs that are unusual. However, 'unusual' has a time context. What may be unusual in a 100-year period or even a millennium can become not uncommon during a one-million-year time span. A great storm, rare over even 1000 years, always happens more than once over a one-million-year period. The chances of the Earth being hit by a big asteroid are infinitesimal, but it has happened and will happen again, at least once every 200 million years. Time has such a smoothing effect in geology that even fossilization becomes common, and it is this, of course, that feeds the stratigraphic column. We have to remember, though, that we are looking at the past through a record of the remarkable and we cannot apply our own perception of the common, with its 80-year timespan, to the geological perception of common, with its timespan of millions of years. But even in geology, there are still unique events.

This is true for the Achanarras fish bed, which, in terms of the entire Orcadian succession, is unusual. Even though fish beds occur throughout the Middle Old Red Sandstone, the same bed as at Achanarras can be recognized in Orkney and right around the present Moray Firth coast. It is unusually thick at two metres, most other beds being only two centimetres or so. A detailed analysis shows that at Achanarras, as the bed accumulated, so the species of fish changed, even though it is supposed by some to have taken no more than 4000 years to accumulate. The bed clearly represents a special time in the development of the lake and the fish. Five fish genera dominate the fossils while another nine are present but rare. The most common fish are *Dipterus, Palaeospondylus, Mesacanthus, Coccosteus* and *Pterichthyodes* although *Osteolepis, Diplopterus,* and *Glytolepis* also occur. These names are frightful and one sentence from Hugh Miller explains it all:

> ...I was led to confound the Osteolepis *with the* Diplopterus, *and to regard the* Glytolepis *as the* Osteolepis...

He wrote in all sincerity, meaning that even to a specialist some of the fish look pretty well alike, especially when looking at the black, fossilized, squashed carcasses! Try identifying your favourite cat, squashed on a busy motorway! However, let us first describe *Dipterus valenciennesi* (Sedgwick and Murchison), mainly because it is the commonest fossil at Achanarras and the one most often found there by fossil hunters. It illustrates the typical characteristics of these Devonian fish. The name, of course, follows the usual biological system inherited from the Swede Carl Linnaeus; the first name is the genus, the second the species, and in brackets the original author(s) who described it.

Fossils of *Dipterus valenciennesi* (**Figure 3.6**) are on average about 20cm long, the size of a large sardine. *Dipterus* is a dipnoan, of the

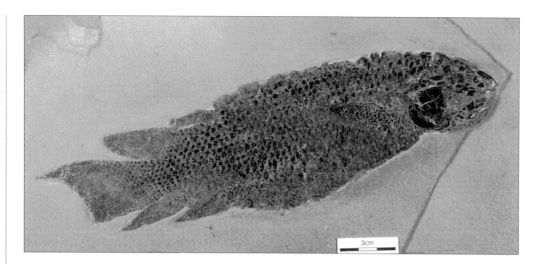

Figure 3.6
Dipterus valenciennesi,
one of the commonest
fossils from
Achanarras (courtesy
of Nigel Trewin).

order *Dipnoi* (meaning double breather), which means it is a lungfish with lungs as a breathing apparatus in addition to gills; species of lungfish still live today in the southern hemisphere. Lungfish belong to the family *Sarcopterygii*, that is, they are lobe-finned, bony fish, as opposed to ray-finned, bony fish which are the most abundant today. There is a second group of bony, lobe-finned fish: the *Crossopterygii* (also family *Sarcopterygii*), which differ from the *Dipnoi* in fin layout, head structure and teeth. We will look at these fish in more detail later in this chapter, since they are on the evolutionary line to amphibians and eventually to us. The overall body shape of *Dipterus* was rounded like that of a trout, more so vertically than laterally, so that it should have been able to swim moderately well and have been able to chase prey. It had a horizontally flat head, which it seems to have used like a vacuum cleaner, sucking up worms and suchlike along the lake bottom in shallow water. But *Dipterus* also had tooth plates, the style of which suggest that it was omnivorous, the strong jaws indicating that it could eat shellfish, snails and crabs, as do the modern lungfish. It clearly cannot have lived on the quiet lake bottom in deep water where it was fossilized, there being neither food nor oxygen. The head of *Dipterus valenciennesi* is made of a complex of bony plates and the body is covered with distinctive, primitive scales that, even then, were made of calcium phosphate like our teeth. In Achanarras, all these hard parts are fossilized in black, phosphate-rich material and stand out from the dull greenish grey of the rock itself. Even single scales, because of their black, shiny colour, are easy to pick out. The tail of *Dipterus valenciennesi* is heterocercal, a distinction that marks these ancient fish and means that the body continues right into, and is part of, the tail, whereas in all modern fish the tail is a simple, finned addition to the body (homocercal). *Dipterus* has two body-centred fins on its back (dorsal) quite near the tail and five belly (ventral) fins, one (the anal) equally near its tail and, importantly, two paired fins, one just behind its head and the other a little forward of its tail (pectoral and pelvic). These will become the legs of the future, as I will illustrate later in the chapter.

The many complete fossils of *Dipterus valenciennesi* in the Achanarras beds show that the dead fish would fall to the lake bottom upside-down, presumably because of air or gas in the stomach or lung, the flat head being preserved in this fall position. The body, however, as it rotted in the perfectly stagnant water on the lake bottom, would detach from the head and fall sideways. What is now preserved on the flat layers is a profile of the carcass often with the writhing, half-moon, upwards curve characteristic of *rigor mortis*. The anguish of death is fossilized with the fish, the stomach even appearing bloated from early, rancid decay. It is common to find single scales scattered around the side-on carcass of a *Dipterus* for several centimetres, as though when it finally fell, probably stirring up a small cloud of fine mud, loose items fell off the rotting body like flower petals in a gust of wind. *Dipterus* is not the only species to show fossilized death throes; each fish has left distinctive marks of its own way of death. Such a theatrical gesture attracted Hugh Miller to these fish and prompted him to write:

Geology of all the sciences, addresses itself most powerfully to the imagination; and hence one main cause of the interest which it excites.

Figure 3.7
The tiny *Palaeospondylus* with the typical body curve as if in the anguish of death, with an additional lengthwise twist (courtesy of Nigel Trewin).

Palaeospondylus gunni (**Figure 3.7**) was a funny little eel-like animal, presumed fish, of which there are many examples in a fossil collection made in the 1890s by Alexander and Marcus Gunn, hence the species name. It was the first fossil found on an initial visit to Achanarras despite its small size, being usually only about 2cm long; this one was 1.5cm, the same as a young aquarium fish. Although this one was alone, it can occur in huge numbers on particular bedding planes, as though a whole shoal had died at once and

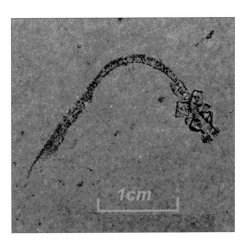

was then sorted by a light wind or drifting current. It is quite unlike the other fossils and may in fact be the young form of something bigger, but no-one knows. Like *Dipterus*, this poor fish also died in anguish and is always found in the curl of death, as was this one, the body even showing an additional lengthwise twist of pain between the head and the tail, like a slice of used lemon. The fossilized death throes did not escape Hugh Miller, who observed:

> *The attitude is one of danger and alarm; and it is a curious fact, to which afterwards I shall have occasion to advert, that in this attitude nine tenths of the Prerichthyes of the Lower Old Red Sandstone are [...] to be found.*

The death of one well-known specimen of a fish called *Glytolepis paucidens* (Agassiz) does not, however, provoke sensitive feelings. It choked to death on its last meal as it was trying to swallow one of its own kind, and is famously fossilized with the victim still stuck in its mouth. No need to analyse the diet of this fish; it was obviously a carnivore and a cannibal at that!

A strange fish that still defies explanation, and one of Hugh Miller's favourites, being named after him, is *Pterichthyodes milleri* (**Figure 3.8**). Miller spent eight entire pages of his book *The Old Red Sandstone* describing it, telling stories of how he sent it to Agassiz and Murchison for identification and how the Frenchman Lamarck, whom he calls 'that ingenious foreigner', would have interpreted the fish's 'wings'. For this small, oval fish, strikingly covered with large, wrinkled plates, has two long, thin, very distinctive wings or arms or paddles, the word used depending on your interpretation of them. In the end, Miller gave no better explanation than Lamarck, simply wondering what God gave the fish these appendages for, but not deciding for himself. Modern suggestions have not progressed either, and the animal is simply relegated to a bottom feeder and the extensions called paddles (which they clearly were not).

Figure 3.8
Pterichthyodes milleri (Agassiz), the fish found by Miller and named after him by Agassiz. The use of the appendages is still not known (after Miller, 1873).

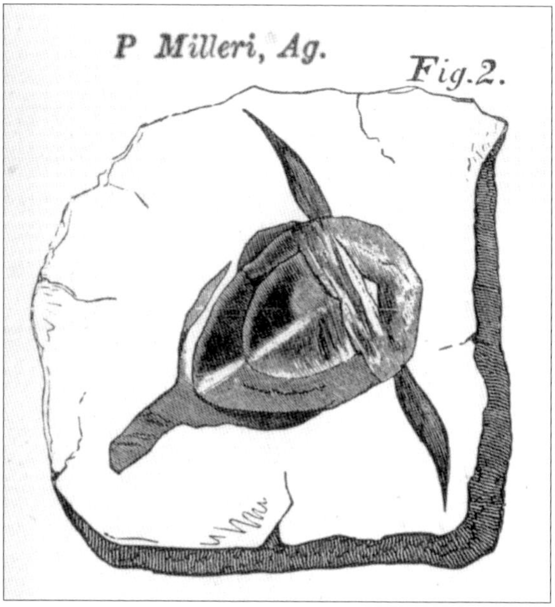

P Milleri, Ag.   Fig.2.

There is a unity of character in every scale, plate and fin, – unity
such as all men of taste have learned to admire in those three
Grecian orders from which the ingenuity of Rome was content to
borrow when it professed to invent, – in the masculine Doric, the
chaste and graceful Ionic, the exquisitely elegant Corinthian

is Hugh Miller's awestruck attitude to 'his' fish. Beautiful as they are
in themselves, the small, black, shiny, fossilized scales and bones
of these very ancient Devonian fish are most exceptional not for
such peculiarities as 'wings' or for their anguished deaths, or even
their Grecian perfection, but for the similarities that they bear to us.
Some of them are the earliest forms of life that display convincing
features that link them to ourselves. They had a primitive backbone
(a notochord), four paired, muscle-operated bony fins (future limbs),
a skull containing a concentration of sensory organs and a linked top
jaw but movable lower jaw with enamelled teeth, two nose holes (nasal
openings) linked to lungs (but gills as well), and blood containing
urea. The way in which these fish gradually evolved into amphibians,
reptiles, early mammals, apes and man has affected us all since
Darwin proposed it, be we religious, scientific or just plain human.
Hugh Miller would have been appalled by the connection. He kept
animals away from man just as he kept the social classes apart: a man
of his times.

Among these fossils of Achanarras are Crossopterygians, the class
of fish considered to have given rise to tetrapods, animals with four
limbs. Uncomfortable as it may be to have fish ancestors, what feels
even more unnatural is that we still have within us many features
already possessed by these ancient fish. When evolution was first
proposed it was difficult to believe, but now that we mostly accept it,
we can observe almost objectively the way that it has taken place. What
is now difficult to believe is not what has changed but what has *not*
changed; nature, we discover, is extremely conservative (parsimonious
is the word anatomists like to use). In truth, what allowed Darwin
(and Wallace) to identify evolution was that many features remain
so recognizable. In many cases, evolution simply means putting on a
wig or a false nose. The animal underneath is still the same and the
Crossopterygian fish of Achanarras already had many of the basic
design features we still use today. Interpreting squashed, 380-million-
year-old fossil fish is a difficult task even for specialists, but by great
good luck and a remarkable chance there are Crossopterygian fish still
alive today. Granted, there is only a single species, but that one species

is enough for us to be able to touch, examine, cut open, put under the microscope and to observe a living fossil: which it truly is. It can even be cooked for a meal. This single species is the coelacanth.

The finding of the coelacanth is a story from science that still holds attention. Luck is there at the start, but then so are the gifted amateur, the dedicated scientist, the politician and the maverick adventurer. It is a tale good to tell because it is so very unscientific. So much so that the story of the discovery of *Latimeria chalumnae* (Smith), the first coelacanth to be recognized, is far more familiar than is its scientific significance. Finding a living fossil is perhaps justification enough for many, but this is not so. This living fossil is not a missing link or some life form which until discovery was considered not to exist; it quite simply is a fossil found alive. The gap between living fish and the most recent coelacanth fossil is more than 75 million years, the last fossil being from the Upper Cretaceous! The first fossil, though, is of course Devonian and 380 million years old. Palaeontologists study fossils while biologists study living things, an unhelpful division which should not persuade either specialist to ignore the other, although they often do. It is the difference between a detective and a reporter. The palaeontologist generally has only hard parts to study, or at least the parts that are fossilized, as we see with the Devonian fishes. It requires detective work to reconstruct the fossil's way of death and its way of life. The biologist simply looks at the forms he is interested in and reports on his observations. The palaeontologist dreams of seeing a fossil alive, and when it comes to dinosaurs, so do many children, which seems to include most adults. The coelacanth is a living, swimming, breeding, 380 million-year-old fossil in the warm waters of the Indian Ocean. A palaeontologist's dream come true, and far more remarkable than a living dinosaur since it is more than five times as old. It is far more important for having some of the oldest vertebrate DNA and for being one of our earliest, but living ancestors. Thank evolution for the coelacanth.

When the first *Latimeria* (coelacanth) was discovered in 1938, mixed up in a random catch of Indian Ocean sharks discarded on the deck of a boat in East London docks on the coast of South Africa, it was quite by luck. So lucky, in fact, that even with a long and dedicated search, it took another 14 years to find the next fish. It was found in the Comoros Islands near Madagascar, 1000km away from the first find, and from where many more specimens have eventually come, and where the fish actually lives. Because the fish dies when brought to the surface, all the early specimens were simply dissected

and their anatomy copiously described. The French biologist Jacques Millot took 18 years and three large volumes to do just this (the Comoros were a French colony at the time, and the fish considered to be French property). It was only in 1986, nearly 50 years after the famous first find, in a small, home-made submersible built by Hans Fricke, that the German Jürgen Schauer actually saw the first coelacanth alive and swimming. He was at a depth of around 200m in amongst the sea caves on the flanks of one of the volcanic Comoros islands. We can now live a palaeontologist's dream and look at the living coelacanth, at its remarkable biology and lifestyle, bearing in mind that we have already looked at its 380-million-year-old squashed, shiny black, fossil ancestors. We would like to know how, in those 380 million years, the tetrapods (four-limbed vertebrates) were able to evolve from them, rise out of the sea and walk on to the land. We do not agree with Hugh Miller, who suggests:

> *There is no getting rid of a miracle in this case, – no alternative between creation and metamorphosis.*

He could only see creation; we prefer metamorphosis and call it evolution.

The most emotive link between fish, tetrapods, and us, is four legs. A walking fish is an incongruous image worthy of a nonsense poem. Seeing a crocodile walk on land, waggling its whole body ridiculously from side to side just to be able to move its stubby legs a little, is to wonder why such locomotion is ever useful to them. The crocodile is at home in water where, even though it does have legs, its tail powers its movement, just as it did for its fish ancestors.

Figure 3.9
Latimeria chalumnae (Smith), the coelacanth (photo ©Atypeek Dsgn. Shutterstock).

That has always been the puzzle. Being so well adapted to water, why did fish develop legs in the first place? However, leg structure clearly links us to the amphibians, and them to the Crossopterygian Devonian fish, the ancestors of the living coelacanth. The arrangement of bones in our hands and feet is extremely odd, but incredibly it is repeated in all animals with limbs, and its persistence over hundreds of millions of years is a major puzzle. A single bone attached to the body (femur in the leg and humerus in the arm), then two bones extending the limb (tibia and fibula in the leg, radius and ulna in the arm), followed by a complex of bones in the wrist (carpal bones) and ankle (tarsal bones), before a multiple digit termination, mainly five (fingers and toes), is the structure in 350-million-year-old amphibians, 150-million-year-old dinosaurs, and us. Of course there are and have been modifications, but they are only that: modifications. The basic design remains the same: kangaroo legs, human arms, bats' wings, horses' legs, cats' paws are all based on the same pattern. That this design was created only once, over 380 million years ago, and has remained the same ever since is startling enough, but there is more. Digit number has been five from very early on, and the rule applied even to the dinosaurs. Digits are lost but not multiplied, even though six-fingered human hands are possible. Five is so much the rule that there was a considerable palaeontological fuss when amphibians with seven and even eight digits were found in the Upper Devonian of East Greenland. This is strange enough, but what certainly needs reflection is that all limbs have the same number; the eight-toed front legs on Acanthostega gunnari, one of the Greenland Devonian finds, are linked with eight-toed back legs. It is clear that in evolutionary terms, digits appear at the same time on hands and on feet and in the same number; there are no Devonian tetrapods with fingers but no toes, or different digit counts. Something has always linked limbs together, and still does, in evolution and also in function, limb movement being synchronized even in the poor waddling crocodile.

And what about the living coelacanth? The fish does not walk, even though JLB Smith who first described it called it 'Old Fourlegs' and thought that it should and would. It swims, and one of its preferred positions, as observed by Hans Fricke from his semi-submersible, seems to be on its head. The coelacanth is significant, though, because it has two sets of paired, bony fins, just as its Devonian ancestors did. When the coelacanth swims, the bony fins can be seen to move in a co-ordinated sequence like tetrapod limbs: the front left and hind right fins making an identical front-to-back figure-of-eight movement at

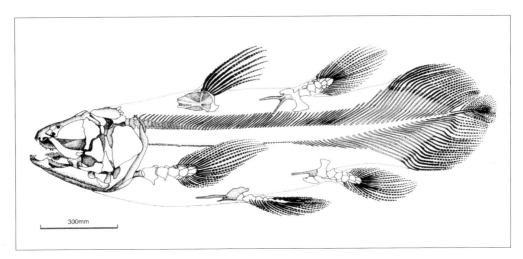

300mm

Figure 3.10
A skeletal diagram of the coelacanth showing the ventral (underbelly) fins which move with the same alternate rhythm as our own arms and legs. The fish does not walk but the fins evolved into arms and legs (after Forey 1998, courtesy of the Natural History Museum, London).

the same time, linked to the front right and hind left fins in a similar, but counter movement, the full sequence being rhythmically repeated. Watching a splendid military parade with all the soldiers putting right feet forward and left arms back in precise unison does not immediately suggest the movements of a coelacanth, but the similarity is there. Such sophisticated movements by the paired fins are possible only because they are made up of a jointed complex of cartilaginous bones, increasing in number towards the extremity. 'Old Fourlegs' may not walk (its Devonian ancestors could have), but the modification of the bones of its fins did lead to the development of the tetrapod limb. The coelacanth fin bones are not actually the closest to the tetrapod's; instead a Rhipidistid fish (the other branch of the Crossopterygians), called Eusthenopteron, has individual lobe fin bones startlingly similar in detail to those of a tetrapod limb. That the coelacanth has complex, manoeuvrable fins, unlike almost all modern ray-finned fish, is what makes it so recognizable as a living fossil and such a convincing ancestor to tetrapod and human.

In 1866 a young German, Ernst Haeckel, published his discovery that the evolutionary changes then being proposed by Darwin were actually re-enacted during the early stages of embryo growth, or so he thought. He saw, for example, that in the human embryo, limbs first appear as buds at about 24 days, while limb bones themselves only begin to form progressively a week later, shoulder to finger. Limbs do not grow directly; the embryo goes through the changes from fin to limb that would be expected from evolution. His phrase, 'ontogeny' (individual, especially early embryo, development) recapitulates 'phylogeny' (evolutionary development of the group), sounds very learned and is memorized by every geology student. The theory was

Figure 3.11
The role of Hox
genes. *Drosophila
melanogaster*, the
fruit fly with fully
formed legs where
antennae should
be. Hox genes have
played an important
role in vertebrate limb
development (courtesy
of Visuals Unlimited).

initially extremely successful, but then became totally discredited when so many developing embryos did not show Haeckel's evolutionary patterns. But as is so often the case, old theories never die but are eventually re-examined in a new light, and in the last 10 years this has happened with Haeckel's ideas. The consensus today is that, indeed, a significant part of some evolutionary changes are re-enacted in embryo development. This means some evolutionary changes can be analysed and eventually understood by observing living embryos. What has been discovered is that amongst the genes that code for particular features, there are genes that act as switches, dictating when, and consequently where, things should happen. Limb structures have been constant and linked for millions of years because their gene switches have stayed the same. They are very ancient.

There are strange freaks in the animal world in which limbs become mixed up: mutant moths with wings as legs, flies with feet on the ends of their antennae. This is puzzling because the wings are still wings, the feet, still feet. Geneticists now explain such errors in terms of gene switches, the errors being made by the switches and not by the genes responsible for the morphology. Wing genes switched on during leg growing time grow as wings. These freaks are only interesting for what they indicate about normal development, especially for limbs. Fins and limbs begin in the same way in an embryo, as buds or swellings. The subsequent differences can be followed in the timing and behaviour of Homeotic or Hox genes, as they are called, which control the development of protruding body parts. Hox genes may control development, but the activity of these genes themselves is controlled by switches that are present in the chromosome but are not actually genes. The switches allow the Hox genes to be active or dormant. In the common ray-finned fishes, Hox genes are switched on early and slow down any further change; multiple rays are formed on the bud. In lobe-finned fishes, like the coelacanth, the Hox switch is delayed, which allows bones to develop but not digits. Finally, in limbs, the switch is even more delayed and digits develop on the bones, the number of digits depending on when the switch acts on the Hox genes; the longer the delay the greater the number. This explanation of limb development requires that the growth axis is not simply from shoulder to fingertip or from a single bone to many. It has to pass through only one of the paired bones (e.g. tibia/fibula, radius/ulna) and then into the wrist bones, implying that it is hooked. The digits are therefore to the side (to the rear = posterior) of the growth axis. This also explains why there can be more than five digits and why digits have never further divided; they are branches off the main growth axis and not on the main axis itself.

Our limb link to the coelacanth fish is easy enough for us all to understand. But for the palaeontologist a more favoured link is through cranial or head development. Bone plates in the heads of fishes, amphibians, reptiles and mammals show changes that can be followed progressively through their evolution. There are holes for eyes and for breathing (an indication of lungs) and each bone plate becomes slightly modified in outline and location as the head changes shape. Importantly, of course, head bones are prime fossil material, and if some part of an animal is going to be fossilized, especially of a land animal, then it is going to be the skull. It is quite uncanny, though, to realize that the human skull is initially made up of the many 'plates' inherited from Devonian fishes and that these are visible in the developing foetus before they anneal in the adult. Following the detail of skull bone evolution is best left to palaeontologists, though, as it gets rather complex when each bone has its own name and there are a large and bewildering number of them. Notwithstanding, the uncompromising set of pictures put together in the 1950s by the famous biologist-palaeontologist Alfred Romer can be understood by us all (Figure 3.12).

Surprising as these limited changes in the skeletal design of vertebrates may be, the preservation of the body functions that allow life in salt water, fresh water or air, over hundreds of millions of years, is even more remarkable. There is still debate over whether tetrapods came from marine or fresh water, although all living amphibians are non-marine, implying that their ancestors were as well. As usual, the specialists do not agree. To examine this, it is useful to limit the argument to body functions, which of course are not fossilized. A wild, silver-sided salmon caught in a spring run is beautiful to eat, the flesh crisp and light pink. It is also difficult to catch and full of energy. A late-run fish which has had to wait most of the summer in a brackish estuary before being able to swim up the river is black-sided, has little energy and is horrid to eat, the flesh soggy, sponge-like and tasteless. Like all bony fish, the salmon has to maintain a constant salinity in its body fluids regardless of the fluid that surrounds it. Like them all, its body fluids have a salinity that is neither that of seawater nor

Figure 3.12
Alfred Romer's
evocative illustration
of cranial evolution
from the bony fish
(top left) to humans
(bottom right) (after
Romer 1941, courtesy
of University of
Chicago Press).

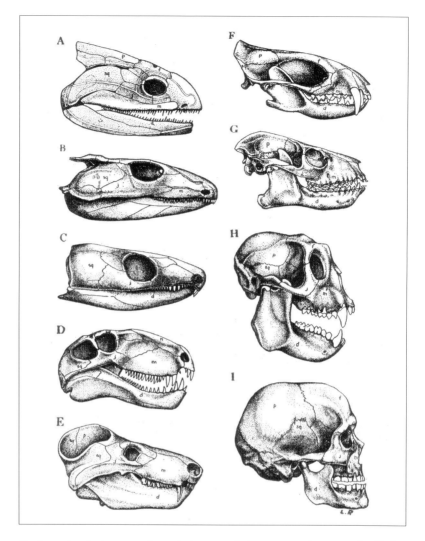

fresh water but somewhere between; it is constantly out of salinity balance. In seawater it drinks a lot of water to make up for the osmotic imbalance, and then has to eliminate the excess salts: in fresh water it is the reverse, and it has a tendency to dehydrate and take up any salts it can. As far as the eater is concerned, salmon are best suited to seawater.

Fish have lived in the oceans for hundreds of millions of years, but all of them still have this salinity problem; they are not designed for seawater. Combined with this hundred-million-year-old problem is another, equally ancient one; the food that they eat produces far too much nitrogen for their body system. Although they process the excess nitrogen into ammonia, it is poisonous and must be eliminated from their body fluids and blood. Modern, bony, ray-finned fish (i.e. most teleosts) do this by flushing lots of water through their gills, which carries away the very soluble ammonia.

This is what salmon do but coelacanths do not; they convert the ammonia to urea, a compound which is not soluble in water and so can be flushed away whenever necessary, and not continually. Sharks use the urea system, so do tetrapods and so do we; our urine is full of it. If you are going to live in air and not water, being able to make urea is essential, since the body system does not require to be continually flushed with water, an impossibility in air! Therefore two basic body fluid systems have been invented. The first, in which great quantities of water are used to eliminate excess ammonia, is used by seawater-dwelling fish, also incidentally allowing them to maintain body fluid salinity. The second, in which ammonia is converted to insoluble urea, is used by lungfish and coelacanths, since it allows great independence for maintaining body fluid salinity, and is the system which has been adopted by the tetrapods in parallel with air breathing. Both systems are over 380 million years old and we still use the second. The answer to the question of a freshwater origin for the tetrapods would appear to be intuitive. The living coelacanth has been used as evidence that the lobe-finned fish are normally marine and not freshwater dwellers. Clearly, as the only living fossil, the coelacanth, like a football star, will be given far more importance than it deserves. Its air bladder full of fat, however, is a modification for survival in the deep water where it now lives, and not an original feature. The fatty bladder and the oily liver and flesh give the coelacanth neutral buoyancy in its present habitat. In its ancestors it would surely have been an air bladder, as it is in the living, exclusively freshwater lungfish. The fact

Figure 3.13
The body system of the Atlantic salmon (*Salmo salar*) requires constant flushing by large amounts of water because its metabolism is out of equilibrium with both fresh water and seawater, as it is for most teleost fish. The coelacanth has a different, urea-based body system which it passed on to the amphibians and to us, and enables life on land (photo ©Chanonry, Shutterstock).

that all amphibians are non-marine should surely count as important, as should the fact that the freshwater Orcadian Lake was full of lobe-finned fish.

When Hugh Miller, who died before the publication of Darwin's *Origin of Species* in 1859, attempted to explain why the Devonian fish died out and were not found in younger rocks, his thinking was typically far-ranging, imaginative and logical:

> *...they were created to inhabit a thermal ocean, and died away as it cooled down. Fish of a similar type may now inhabit the seas of Venus, or even of Jupiter, which, from its enormous bulk, though greatly more distant from the sun than our Earth, may still powerfully retain the internal heat.*

Lovely images. With evolution, we think that we can explain the disappearance of the Devonian fish, but perhaps we are looking in the wrong direction, just as Miller was. Today we tend to concentrate on the changes that evolution brings. The story of Achanarras and the Comoros is one to suggest that we should do the reverse and note what has not changed: and there is a great deal. Science fiction films and television have hit on the technique of creating strange beings with odd heads and curious features, but they are all built around the human body so that they can walk around and make a contribution to the story. Evolution, it seems, has done just this; it has taken basic designs and modified them without upsetting what was underneath. Granted, evolution takes a long time, but this makes the lack of fundamental change even more remarkable.

This observation that basic designs are being preserved during evolution has recently been put into stunning perspective by molecular biologists. Darwin, and all palaeontologists, follow evolution using form and shape. This is how we described the changes from coelacanths to humans – as progressive changes in limb design or cranial shape. Molecular biologists, however, look at evolution in quite a different way: they study genes. What they have found is that despite the obvious differences in morphology between, for example, flies, mice and ourselves, there are remarkable similarities in gene sequences. Darwin was careful never to hurt even a fly: perhaps he was subconsciously aware of this similarity.

The chromosomes of the fruit fly *Drosophila melanogaster* have been studied for many years. Building on these studies it is now possible to understand the early development of the fly in terms of its Hox genes (the ones that control the development of protruding body parts). The fly has eight Hox genes grouped in two clusters along one chromosome. One cluster controls the genes that affect the development of the body parts of the front half of the fly, the other cluster controls those of the rear. The Hox genes occur in the same relative order along the chromosome as the body parts on the adult fly that they affect. Incredibly, researchers have found that the same Hox genes in the same order control the development in mice, elephants and even humans. In other words, fundamental growth development of possibly all complex life forms is controlled by the same Hox gene sequences: in effect the same genes. As Sean B. Carroll writes in his recent book:

> *The discovery that the same sets of genes control the formation and pattern of body regions and body parts with similar functions (but very different designs) in insects, vertebrates, and other animals has forced a complete re-thinking of animal history, the origins of structures and the nature of diversity.*

Biologists now contend that bacteria, insects and vertebrates all share the same set of Hox genes, and in the same order on the relevant chromosome. So ancient are these patterns that they probably precede the Cambrian Explosion and have been in use for more than 540Ma – a discovery with very far-reaching consequences.

We can now explain the conservatism underneath evolution as the result of the constant re-use of basic genetic materials. We know that genes were in use 380 million years ago (and much further back) and that their information was held coded in DNA, just as it is today. When we analyse the DNA of the coelacanth we feel comfortable in assuming that similar DNA existed in the fish of Achanarras. Molecular biologists now suggest that some genes are actually even the same. Anyone who has built with Lego will know that it is very adaptable, but that there is always a basic pattern to any Lego construction. At Legoland, it may bear a likeness to Buckingham Palace, but the toy construction is still instantly recognizable as Lego. Genes are the same; they may be very adaptable but there are limits to what they can do, and just like pieces of Lego, they can be connected in different ways but the style of construction is always the same. We often wonder at the similarities between what are otherwise very different animals. Why, for example, is the marsupial Tasmanian wolf so similar in body build to the North European wolf? The conclusion is now that they are built with the same genes. This applies to animals, insects and even plants, one form being a genetically modified (GM) version of the other. What we learn from the coelacanth is that evolution has done a very imaginative job in creating tetrapods out of a basic fish design, although the possibilities were always going to be limited. We can only look on and admire: after all, we are talking about ourselves!

From the shiny black fossil fish of the Achanarras quarry to the warm volcanic Comoros islands of the Indian Ocean is a distance of 380 million years and many surprises. But Hugh Miller must end this chapter, just as he started it:

> *The wonders of geology exercise every faculty of the mind, - reason, memory, imagination; and though we cannot put our fossils to question, it is something to be so aroused as to be made to put questions to one's self.*

Now what questions would those be, nice Mr Miller?

Further Reading

Books, Pamphlets

Carroll, Sean B. 2005. *Endless Forms Most Beautiful. The New Science of Evo Devo*. New York & London: W.W Norton & Co. Inc. pp.350. ISBN 0 393 06016 0

Dineley, D.L. 1994. *Aspects of a Stratigraphic System: the Devonian.* London: MacMillan. pp.223. ISBN 0 333 25641 7

Forey, Peter L. 1998. *History of the coelacanth Fishes.* Natural History Museum, London. London: Chapman & Hall. pp.419. ISBN 0 412 78480 7

Gostwick, Martin. 1993. *The Legend of Hugh Miller.* Cromarty Court House Publication, pp.55. ISBN 1 898416 03 6

Gradstein, F.M., Ogg, J.G. *et al.* 2004. *A Geologic Time Scale.* Cambridge: Cambridge University Press. pp.500.

Miller, Hugh.1841. *The Old Red Sandstone.* Edinburgh: William P. Nimmo. 17th Edition (1873). pp.384.

Moy-Thomas, J.A. 1939. *Palaeozoic Fishes.* London: Methuen & Co Ltd. pp.149

Romer, Alfred S. 1933. *Man and the Vertebrates.* Harmondsworth: Penguin Books Pelican edition 1954. 2 Vols.

Radinsky, Leonard B. 1987. *The evolution of Vertebrate Design.* Chicago: University of Chicago Press, pp.188. ISBN 0 226 70236 7

Saxon, J. 1975. *The fossil fish of the North of Scotland.* Thurso: Caithness Books.

Trewin, N.H. (Ed.) 2002. *Geology of Scotland.* Geological Society London. 4th Edition, pp.576. ISBN 1 86239 126 2

Trewin N.H. & Hurst A. (Eds.) 1993. *Geology of East Sutherland and Caithness.* Geological Society of Aberdeen. Edinburgh: Scottish Academic Press. pp.183. ISBN 0 7073 0731 7

Further Reading (continued from page 101)

Weinberg, Samantha. 1999. *A Fish Caught in Time.* London: Fourth Estate. pp.239. ISBN 1 85702 906 2

Winchester, Simon. 2001. *The Map that Changed the World.* London: Penguin Books, pp.338. ISBN 0 140 28039 1

Zimmer, Carl. 1998. *At the Water's Edge.* New York: The Free Press, Simon & Schuster Inc. pp.290. ISBN 0 684 83490 1

Scientific Papers

Ahlberg, P. E. and Milner, A.R. 1994. The origin and early diversification of tetrapods. *Nature.* Vol. 368, pp.507–513.

Shubin, N, Tabin, C and Carroll, S. 1997. Fossils, genes and the evolution of animal limbs. *Nature.* Vol. 388. 14th August, pp.639–648.

Sordino, P., van der Hoeven, F. and Duboule, D. 1995. *Hox* gene expression in teleost fins and the origin of vertebrate digits. *Nature.* Vol. 375. 22nd June, pp.678–681.

Trewin, N.H. 1986. Palaeoecology and sedimentology of the Achanarras fish bed of the Middle Devonian Old Red Sandstone, Scotland. *Transactions of the Royal Society of Edinburgh, Earth Sciences.* Vol. 77. pp.21–46.

Zhu, M., Yu, X. and Janvier, P. 1999. A primitive fossil fish sheds light on the origin of bony fishes. *Nature.* Vol. 397.18th Feb, pp.607–610

Chapter 4   **VOLCANO**

The Tertiary volcanic province and the Atlantic opening

A long trip even for humans to make – 24,000 kilometres each year to Brazil and back – but astounding for an unremarkable-looking, medium-sized, black, white and grey seabird. That is what the Manx shearwater does every year of its long, possibly 30-year life. Not only that: each year it returns with the same mate, to the same burrow, on the same small northwest Scottish island of Rum. Are these birds Scottish or are they Brazilian? They live in the south but by birth are Scottish: so Scottish they are. In the soft earth between the dark rock ledges and boulders in the high places of Hallival, Askival and Trollaval, the so-called Rum Cuillin, the birds have scratched deep holes in which they lay their single egg before crawling inside to incubate it by turns, the free bird stuffing itself at sea ready for its feedless period of sitting. By day not one bird is seen, even though there are 60,000 pairs on the island, but their presence is evident from the brilliant 'shearwater greens', patches where rich grasses grow on the fertility of centuries of birds' droppings. In darkness late at night, however, the absent birds come back from the sea, wheeling and squealing in the night as the brooding birds respond from their burrows, the volume of their eerie, mournful cries echoing through the black silhouettes of the mountain peaks as if they were trying to be free of some baleful spell. Even with such a life, though, Manx shearwaters still have a sense of humour. They feed their single offspring so much seafood that it eventually becomes too

Figure 4.1
The Rum Cuillin, a worn down volcano, seen from the south. The edge of a fossil lava flow on the Isle of Eigg is seen in the foreground.

fat to get out of the burrow where it was born. When the parents see this they lose interest and fly off, leaving the over-weaned fledgling to itself. The poor abandoned bird loses weight quite rapidly, and after a few days is thin enough to scramble out, take flight and go and look for its own first meal. It is tempting to draw human parallels.

Rum may be remarkable for its Manx shearwater population, but it is remarkable for more. The island is an ancient, stripped down, eroding volcano, the burrows of the shearwater being dug into its crumbling, weathering heart. It is old, of course, but 60.5 million years ago in the early Tertiary, where the birds nest today was the searingly hot, molten core of a towering 2–3km high volcano set in desolate fields of cinder-black lava. Today, in clear weather, from the rocky 812m (2650ft) peak of Askival, the highest on Rum, you can look south across the choppy sea to the low hills of Ardnamurchan and to the rugged outlines of Mull's Ben More or then north the short distance across the blue water to the dark, impressive peaks of the Skye Cuillins. Each is close but they are all, like Rum, the eroded hearts of 60-million-year-old, early Tertiary volcanoes equally set in the low, slab-like flats of ancient lava fields. A scene once fiery and stupendous but now only cold, a scene of bare rock ribs rotting like ancient shipwrecks. These are the geological monuments of the British Tertiary Volcanic Province, and there are six fossil volcanoes that can be visited today; Rum, Skye, Ardnamurchan, Mull, Arran and St Kilda.

Since the 1960s it has become clear that the volcanoes and lavas that we see in the Western Isles are just a small corner of vast, early Tertiary lava fields that cover an area stretching for 2000km along the continental margin from northern Norway to the Rockall Bank.

This is an area larger than the whole of Great Britain and now called the British Tertiary Volcanic Province. What we see in the Cuillins or on Rum may impress but they are only a tiny part of a huge tectonic–volcanic event caused by the splitting open of the Earth's crust and creation of the North Atlantic Ocean. Yet again, the Highlands were caught up in dramatic Earth upheaval, this time with volcanoes, lava eruptions and continent break-up. As far back as the 1790s, a few geologists realized the volcanic origin of the Tertiary basalts of Skye, Mull and the much-pictured Isle of Staffa, although they did not agree that they were lavas. More of this later. Today, discoveries continue to be made, but now by the oil industry and environmental agencies, who use their sophisticated and expensive geophysical data-gathering techniques to reveal the rocks that exist below the sea, now buried under a kilometre or so of sediment, hundreds of miles offshore. We will return to these modern wonders later, but begin first with what can be experienced on the island of Rum, and a walk into a volcano.

There is only one official place to land on Rum, the small 13 by 14km island entirely owned by SNH since 1957, and that is at Loch Scresort, a timid embayment on the east coast of the island. The Caledonian MacBrayne ferry that plies the passenger route from mainland Mallaig is too large to approach the jetty in the shallow nearshore water, and so waits out in the loch for a small open boat from the island to come precariously alongside. It is the final link to the island. Passengers jump desperately with their belongings into the heaving skiff as if the ferry were about to sink, and then sit tensely on the wooden bench seats while the small boat scurries quickly back to shore. But the system works; people and supplies arrive safely. If there is only one place to land, there is also only one place to stay: Kinloch Castle. The castle is an ugly folly from a previous, Edwardian age, built when many of the unnecessarily rich English spent a good deal of their money in Scotland. It is a rich businessman's dream home but in its realization is a nightmare: a Frankenstein of a building. It makes being rich look rather ridiculous. Kinloch Castle was even the second place in Scotland to have electricity! The servants' quarters of this pretend castle are now a hostel for visitors; the laird's part is a museum, although of what it is uncertain. Pointlessness, perhaps? And so all the paths, to any part of the island, start from Kinloch, and one is tempted to say like 'all roads lead to Rome'. Kinloch is definitely not Rome, though.

Almost all of Rum is bare and treeless now – how much naturally so is uncertain – but because it has been protected from grazing sheep and deer for more than 40 years, Kinloch itself is thick with trees and undergrowth. Leaving the castle and taking the path southwards to Coire Dubh leads directly to the island's highest peaks of Askival and Hallival. Out of Kinloch, coming abruptly onto the bare moor from the preserved enclosure of trees, the Coire Dubh path climbs quickly upwards alongside a bustling, rocky burn. Not all of the island of Rum is volcano. The limit of the volcanic rocks is distinct and marked by vertical faults known as the 'ring faults', which form a neat circle nearly eight kilometres in diameter over the southern half of the

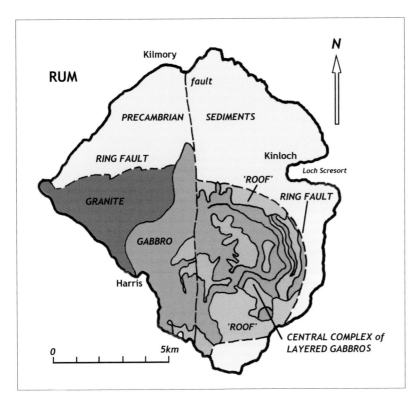

Figure 4.2
Geological map of the Rum volcano, much simplified. Inside the 'ring faults' is the fossil magma chamber (redrawn from Emelius, 1997).

island; this is the boundary of the ancient volcano core. Outside the ring faults are the original, existing rocks (termed country rocks by geologists) through which the volcano forced itself. Here, where the Coire Dubh path begins to climb upwards away from Kinloch, these country rocks are Precambrian, Torridonian shales and sandstones, familiar from the story of Stoer in Chapter 1, and clearly not part of the volcanic intrusion. However, in the dark hills just ahead we are already looking into the volcano, and even here at the bottom of the path we are very close to its edge. In fact, to follow the Coire Dubh path is to walk into the volcano itself.

Generally the Torridonian rocks that are unaffected by the volcano dip gently at 10° to 15° to the west, an effect which can be seen by looking back northwards over the trees of Kinloch and past Loch Sresort, to where there is a clear pattern of dipping rock ribs crossing the low hills. But along the Coire Dubh path, the Torridonian beds dip steeply beneath our feet at 50° or more away from the volcano, forced up by the intrusion of the volcanic neck. As the path climbs and winds upwards ever closer to the volcano's edge, the rocks themselves become broken, shattered, faulted, twisted: all signs of having once been intimately involved with violent volcanic eruption. Then, quite abruptly, the rocks change; bits of partly melted Torridonian are seen mixed with crystalline volcanic rocks, broken up pieces of breccia occur in confused masses, and there are vertical dyke intrusions. These changes mark the circular, vertical, ring faults that are the limit of the volcano, the outside edge of the volcanic chamber. The mixing of the igneous and the sedimentary rocks, a confusion of volcanic and country rocks, is to be expected.

The path still climbs up across the confused outer limits of the volcano, but over a sharp crest the track suddenly disappears into the tall grass of the wide, flat, boggy floor of Coire Dubh. Circling the boggy flat, like a huge amphitheatre, is the strange, stepped face of the corrie wall. It has the feeling of being man-made, not natural, perhaps the terraces of a large but long-deserted quarry. It is, however, quite natural; no human undertaking has carved out the high benches and it is the first view we have of the rocks of the heart of the volcano. The terraces are made of thick layers in the crystalline rock that have been naturally picked out by weathering. They were formed as molten liquids slowly cooled, bench by bench, crystal by crystal. This is a layered gabbro, a dark rock and the crystalline remains of a once searingly hot magma which came up from the Earth's mantle tens of kilometres below, before filling this magma chamber and feeding the ancient Rum volcano. When such a liquid magma cools quickly, as it will when it is erupted onto the surface, it will form dark, finely crystalline, even glassy, basalt lava. However, when it cools more slowly, as happens in an enclosed magma chamber inside the volcano, larger crystals can grow, typical minerals can form, and the rock becomes a gabbro. Layered gabbros are a special case, the magma cooling by stages as the volcano grows. We will describe the geology of layered gabbros in more detail later because it is fascinating and important. It is enough to say here that layering occurs right through the Rum volcano, and is so evident that even individual benches (geological units) are able to be recognized, and are shown on geological maps following the same circular outline as the ring faults. From the floor of Coire Dubh, looking up to the stepped flanks and eventually the peaks of Askival and Hallival, all the rocks ahead are dark, layered gabbros: the pile is over 800m thick. However, before exploring the Rum volcano further, let us introduce some observations on modern volcanoes, and a dash of history.

It is remarkable to realize that only 200 years ago basalt was not associated with volcanoes and it was generally thought that it, and indeed all rocks, had crystallized from water. So taught Professor

Abraham Gottlob Werner (1749–1817) in Freiberg, and for fifty years, from about 1775–1825, he was generally believed. Werner (Chapters 2 and 3) had a complete theory about the Earth and proposed that all rocks formed in a Universal Ocean, from which they were precipitated as crystals in distinct stages. First came primitive granites and gneisses, followed by slates and greywackes and finally limestones, sandstones, chalk and basalt. Columnar basalts were especially satisfying; the columns were obviously large crystals, and the Giant's Causeway rising out of the sea in Northern Ireland, and the Isle of Staffa to the west of Mull, were spectacular examples. No-one really realized at the time that crystals could form from hot melts and not just from water. Werner was called a Neptunist and of course he was wrong, but in his time he persuaded many that he was right because he was a good teacher, his ideas had a charming simplicity and completeness and, importantly, because they conformed to Genesis. Werner did recognize volcanoes (he had to), but considered them as a final, unimportant addition to his master sequence – simply little local irregularities formed once the Universal Ocean had receded. They were heated by coal deposits burning spontaneously just beneath them, which gave rise to lavas, not basalts. Basalts came from the Universal Ocean.

Famously, these ideas were directly opposed by James Hutton (1726–1797) of Edinburgh, who considered the Earth to have a fiery centre from which both granite and basalt came as hot, molten liquids; in other words, they were igneous rocks. Hutton was a Vulcanist: almost the only Vulcanist at the time, he presented his theory in 1785 (considered in Chapter 6). Because his ideas were complex, unconstrained by the Bible and badly expressed in his writings, he was at first supported only by a small group of his Edinburgh friends. Although he was eventually found to have been mainly right about subterranean heat and igneous rocks, Hutton was not quite right about basalt. Like the Neptunists and Werner, he differentiated between basalt and lava, which for him were unrelated. Lava was visibly volcanic, extruded molten, and had only empty vesicles (holes, like in

Figure 4.3
Early sketch by John Clerk of Edlin, 1787, attempting to explain the form of the Isle of Arran (one of the Tertiary volcanoes) using Hutton's ideas of a hot Earth (courtesy of Sir Robert M. Clerk).

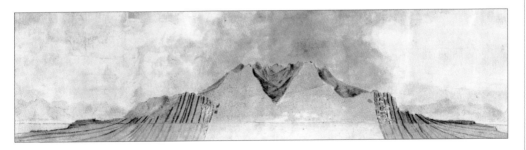

a Swiss Emmental). Basalt, and typically the columnar type as found in the Giant's Causeway or on Staffa, was also igneous but formed below the surface, characteristically containing vesicles filled with minerals, including agate. Today, basalt has become a term of mineral composition with an agreed chemistry. The rock may form as a lava flow on the surface or be injected underground as a dyke or sill, but in whatever manner, was once naturally molten. It is the commonest rock on the Earth's surface.

The debate between the Neptunists and the Vulcanists was long and rude. In 1796 Robert Jameson (1774–1854), a staunch Neptunist of Edinburgh University, sneered:

> *Upon such a basis is the famous Volcanic Theory founded,*
> *which for many years consigned three-fourths of our globe*
> *into the hands of Pluto – until the immortal Werner, from a*
> *careful examination of nature, declared the absurdity of such a*
> *hypothesis (if it can be so called).*

While in 1813 Robert Bakewell sneered back that the Neptunist theory 'is truly ridiculous and will form an amusing page in the future history of science', adding that the theory 'will be preserved from oblivion embalmed in its own absurdity'. He was mainly right, but there were absurdities on both sides and, for example, in 1789 Professor Samuel Witte of Rostock proposed that the pyramids of Egypt were ancient volcanoes because of their shape and their regular jointing, as of course were also those of Peru! Despite the silliness and acrimony, good scientific work was being done, especially by French geologists such as Nicholas Desmarest (1713–1815), Jean-Etienne Guettard (1725–1786) and Barthélemy Faujas de Saint-Fond (1741–1819), who, unlike Werner, were familiar with the active volcanoes of southern Italy, and had also discovered 8000-year-old fossil volcanoes in the Auvergne, central France, associated with columnar basalts. They were the first to present the scientific information necessary to demonstrate that basalt was volcanic. In addition, the increasing number of observations of active volcanoes from around the world from about the 1760s onwards eventually provided all the proof necessary for the igneous, volcanic origin of basalt to be undoubted. By 1830, when Charles Lyell (1797–1875) published his very famous *Principles of Geology*, Werner's ideas of a Universal Ocean had been dropped, both basalt and granite were accepted as igneous rocks, and basalt and lava were viewed similarly and associated with volcanoes. Many ideas about volcanoes were still not correct, but at least they were developing in the right direction: the Earth was now hot.

Modern work on volcanoes is mainly shared between the careful scientific measurements of American scientists, especially on Hawaii, the daring exploits of French volcanologists, who treat their study as an extreme sport, and the emotionally involved Icelanders, who live among them.

There is a separation between these scientists – the daring volcanologists (who get their feet hot, and some even die) – and mainstream geologists (who won't and don't) who only look at outcrops of cold rock, or through microscopes at thin sections of crystalline rock. Most books on the subject still follow one strand or the other. A recent publication on the geology of Rum exists. It is extremely competent and detailed, but there is no mention of a volcano; the word is not used. It has become the emotionally cold 'Rum Central Complex' (to be fair, the word volcano is used on one page out of 170, and does appear in the index, but my point is valid). There is a book called simply *Volcanoes*, in which there is a lot of activity, smoke, lava and earthquakes but no fossil examples. In nature, clearly, an active volcano inevitably becomes fossilized and worn down; but the extinct can still tell us about the living, or of course the other way round, as we shall see. So before we return to Rum, a cold, fossil volcano, we must get our feet well hot on a modern, active one.

Since 1912 scientists have been spying on the volcanoes of Hawaii. It's a nice place to do it – think of the surfing. This is not the cold 'Central Complex' of Rum, but the hot vents of Mauna Loa, Kilauea, Mauna Kea, Kohala and Hualalai, although only the first three of these are active now in 2019. The permanent observatory now run by the United States Geological Survey (USGS) stands on the rim of Kilauea's three by five kilometre crater, peering directly into it. You can even do the same yourself with their webcam. From here, the size of the volcano is monitored: its temperature, the sounds it makes, the vibrations it creates and more. Any Hawaiian eruption is preceded by variations and changes in certain measurements. For example, very sensitive spirit-levels, or in scientific terms tiltmeters, show an outward tilt of the volcano's sides as it fills with hot magma and, like a full stomach, begins to bulge. As eruption takes place the volcano collapses and the tiltmeters begin to tilt inwards. There are also Global Positioning System (GPS) stations that move apart as the volcano swells and get closer as it collapses.

At the same time seismometers – very sensitive noise receivers – sense low-frequency tremors, in effect, a hum caused by the actual movement of the magma, which results in changes in pressure and the release of bubbles of gas. As the tiltmeters and GPS stations indicate, the eruption of lava causes the volcano to change shape, and swarms of shallow earthquakes show that rocks are fracturing and breaking as this happens.

Measurements continue before, during and after eruptions. Sometimes lava is produced and the eruption suddenly ceases, the tiltmeters showing the volcano bulging and then sinking back to its original size. Shape changes can be more complicated than this, however, because it is commonly the case that even as an eruption takes place at the surface, new magma continues to force its way into the volcano from deep below. The eruption from the flank of Kilauea, for example, which started in 1983, still continues – the so-called 56th lava episode as of writing. However, the eruption shows a marked on–off pattern. The tiltmeters indicate that the volcano swells slowly over about a month and then collapses during a rapid extrusion of lava lasting a few hours, perhaps a day, or as in 2018 even longer; this stops and the volcano begins its monthly swell again, and so on. For years it has done this and the lava is still flowing. The explanation is that the inflow of magma from below is constant but slow, much slower than the outflow of the lava from the vent. Each time the lava flows it releases the pressure in the magma chamber but because of the persistent deep magma inflow, it begins to rise again. At the beginning of the 1983 eruption, the lava was generally cooler, but as flow has continued it has become hotter, until it is now at about 1150°C. At this temperature it moves more rapidly, forms different types of lava structure, and has changed composition, becoming richer in magnesium and containing different gases. The early lava had evidently been stored in the volcano for some time while the later, hotter lava has come more quickly from the deep source. It is

Figure 4.4
The tilt record from the summit of Kilauea volcano, Hawaii, from 1956 – 1988. Inflation indicates the slow filling of the shallow lava reservoir inside the volcano. Rapid deflation occurs when the volcano erupts (USGS data, redrawn from Decker & Decker 1981).

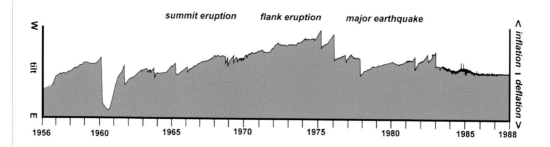

not clear how typical all this behaviour is of volcanoes in general, as Hawaii is built entirely in the ocean and the eruptions are relatively quiet, that is, non-explosive. In contrast, volcanoes on continental landmasses are often explosive, their magma is thicker, containing gases of mainly water vapour and carbon dioxide, which on eruption can explode like a champagne bottle, as happened with Mount St Helens in 1980 and Mount Pinatubo in 1991.

The permanent instruments on Kilauea measure not just the near-surface activity but also what is happening deep below. The Hawaiian volcano is unusual in that it has been built up directly from the deep ocean floor nearly 5000m below sea level, and up to a peak nearly 4000m above the sea, a massive total height of 9000m, higher than Everest. How is it that a magma has the force to build such a huge heap of lava?

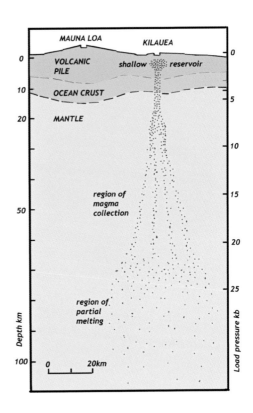

Figure 4.5
The magma source for Hawaii is from deep below the islands. Magma from this source arrives continuously, is stored in the shallow magma reservoir, and then flows out intermittently as lava (redrawn from Holmes, 1993).

Seismic measurements show that small earthquakes related to magma movement do not occur deeper than 50–60km below the summit of the volcano. This is where the magma appears to begin to flow upwards, well below the rigid oceanic crust, which is only 10–12km thick, and into the mantle where temperatures of 1200°C, the melting temperature of basalt, are thought to exist. There is a possibility, however, that the real origin may be even deeper, and that it may only be at 50–60km, where flow in fractures is detected, the rock behaving differently at greater depths. What causes the magma to flow upwards to the surface, though? The answer is the density. Once it becomes liquid, magma has a lower density than the surrounding solid rock and, like oil in water, flows naturally upwards: it floats. To squeeze lava 4km above sea level to the summit of Kilauea, considering the average densities of crust and liquid magma, the supporting magma column must extend at least 57km downwards. This simple calculation and the seismic evidence match nicely: magma begins to flow upwards from a depth of 60km.

But the activity in this deep magma source bears no relationship to the activity measured at the surface or to the changing shape of the volcano detected by the tiltmeters. There is clearly a secondary storage

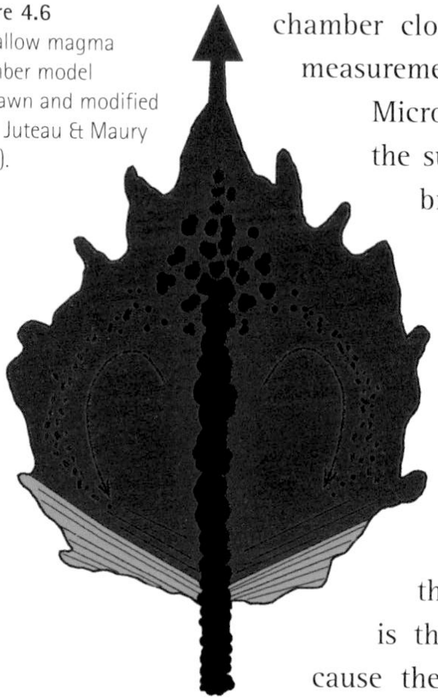

Figure 4.6
A shallow magma
chamber model
(redrawn and modified
from Juteau & Maury
1997).

chamber close below the volcano's surface, and geophysical measurements tell us where it is.

Microearthquakes, or very small shocks detected below the summit of Kilauea, are common and are caused by brittle rock breaking or fracturing as the volcano bulges or collapses – those changes in shape being detected. However, there is an area where no fracture microearthquakes occur, a 'quiet zone', where the rocks do not break, and for the simple reason that they are molten, not solid. In other words, the region of no micro-earthquakes outlines a shallow liquid magma chamber that occurs between three and six kilometres below the volcano crater and is two kilometres across. It is the changes in this shallow magma reservoir that cause the bulging or shrinking measured by the surface tiltmeters. So the volcano has a deep source 60km below and a shallow magma chamber 3000m beneath the crater from which lava is regularly vented. Eruption can only take place if the shallow chamber is replenished from the deep magma source; a toilet will only flush if the cistern is filled from the permanent water supply. Such 'volcanic plumbing' is expected to be fairly typical. Fossil volcanoes, though, will tell us if this suggestion is true, and Rum is an especially useful example.

The obvious characteristic of the fossil volcano of Rum and the other Tertiary centres of Skye, Mull and Ardnamurchan is their circular or slightly oval outline. Make the short climb into the hills above Mingarry Castle at the tip of the Ardnamurchan peninsula and you will be surrounded by clear, circular rings of rock and feel uncomfortably in the presence of something inhuman and unlikely, like a crop circle. From the air, the shape of the Ardnamurchan centre is unmistakeable, and it is surprising to see such a perfect circle. Without thinking, what we expect to see of a fossil volcano is something round or nearly so; Rum certainly obliges. It is easily forgotten that as recently as the 1940s, all the round craters on the moon were thought to be volcanoes, and when the American Ralph Baldwin suggested in 1941 that they were in fact meteorite craters, his paper was rejected by all the 'serious' journals and only published eventually in *Popular Astronomy*. It is now, of course, the opposite and they are all meteor craters, the moon's surface having been pulverized to powder because so many meteorites have hit it over billions of years (see Chapter 7 if this interests you). But on the Earth's surface, round is almost always volcanic, and however deeply eroded a volcano may be, it is still round. This is interesting as it means that a circular pipe or neck or vent extends well below each volcano for thousands of feet. We can therefore examine the Rum fossil for solid facts to relate to the measurements on Hawaii. What is now visible on Rum is only detectable by remote geophysical measurements in any active volcano. We can now go back to Rum and the amphitheatre of Coire Dubh to explore the solid, fossilized magma.

From the floor of Coire Dubh we are faced by the wide, stepped amphitheatre of layered gabbros, already introduced. However, to the left side of the corrie and also to the right are hills with light-coloured rocks, quite different in character to the gabbros. These rocks are a mixture of Torridonian sediments, volcanic lavas and crystalline intrusions, all dipping steeply northwestwards, away from the centre of the volcano, even though they are inside the boundary ring faults, inside the volcano, as it were. Remarkably, these are patches of the original 'roof' of the magma chamber, a vaulted lid that originally covered the molten magma, covered what was to eventually become the layered gabbros. There is a similar piece of roof on the other side of the island, equally inside the ring faults, which, as would be expected, here dips southeastwards away from the centre of the volcano. The most impressive view of this piece of roof is on Beinn nan Stac, just above the Dibidil path on this southeast part of the island. It is well worth the visit.

The coastal path from Kinloch to the old settlement of Dibidil, near the southern tip of Rum, skirts around the eastern edge of the volcano and passes along the foot of the patch of roof. Even in the frequent

Figure 4.7
A section through the magma chamber and the 'roof' that covered it. Southern area of Rum, section passing through Beinn nan Stac and Askival (after Emelius, 1997 courtesy of British Geological Survey).

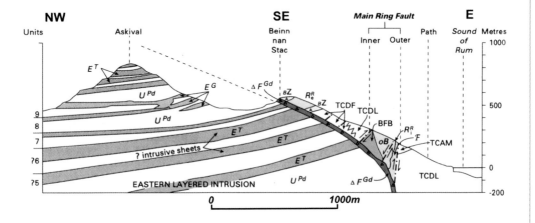

Atlantic mist and rain it is a fascinating visit, and the eventual steep climb is exhausting but rewarding. A north-facing cliff, which marks the northern flank of Beinn nan Stac, is in fact the edge of the magma roof. The roof is formed of broken, baked and fractured Torridonian sandstones, the deep fractures creating an unstable, indented, vertical face with frequently falling rocks. Below the cliff, in steep grassy slopes, are the layered gabbros, the differently weathered layers forming the now familiar giant steps. The layering follows the contours of the hill and clearly passes underneath the cliff, which means below the roof. The layers can be individually traced going under the roof fragment where we stand and coming out on the other side. But if we follow the roof itself, it gets rapidly higher as we climb up towards the peaks of Askival and Hallival. If this slope is projected upwards, it is found to pass over both peaks, meaning that it originally covered all the layered gabbros. It is clear that the Rum volcano is presently eroded down to the level of the roof of the original shallow magma chamber. It is tempting to assume that when the Rum volcano was active, this was the top of the 'quiet zone' detected below Kilauea on Hawaii. Some calculations will show that this is possible.

The volcano on Rum was the first in the area and is older than the Cuillins, Mull and Ardnamurchan volcanic centres. Only the lavas of Eigg and Muck, dated at 61–62Ma and about 300m thick today, existed before the Rum volcano, which is itself dated at 60.5Ma (these are all Palaeocene ages, that is, between 56Ma and 65Ma). The surface onto which the Rum volcano was built would have been the top of these early lavas. If we want to calculate the height to which a volcano can build, Archimedes' law is used, being based on the height of the magma column and the difference in density between the magma and the surrounding rock. This was the method used for the Hawaiian volcanoes. If the density difference is 5% (a typical figure), the maximum volcano height is one-twentieth of the depth to the source.

For Rum, considered to have been sourced from a depth of 60km, this means the volcano could have reached a height of 3000m. Adding together the thickness of the pre-Rum lavas (on Eigg 300m) and the height of present outcrop (Askival 800m), means that originally there could have been 2000m more volcano over the magma chamber roof than we see today. Such a shallow cover is quite surprising, but the equivalent magma chamber in Hawaii, the top of the 'quiet zone', is only 3000m below the surface, and as the volcano there is three times bigger than Rum would have been, the figures are comparable. From its 3000m summit, the active Rum volcano would have spewed its lava over a wide distance and covered the present sites of the Isles of Canna and Eigg with quite a thickness. Some flows would have reached the southern tip of the Cuillins on Skye just 30km away. Today, of course, this huge, impressive pile has all but disappeared.

On Skye and Canna, the thick lavas that we see now have ages between 58.9Ma and 60.5Ma and so were being erupted at the same time as the 3000m volcano on Rum was dying and being eroded away. Huge water-worn boulders of lava up to two metres across were torn from the flanks of the volcano in fast-flowing torrents and then dumped into the surrounding lava fields. This debris is now found fossilized between the flows. On Rum itself, the debris is found as coarse conglomerates filling fossil valleys eroded into the remnants of the once towering volcano. The strange peak of Orval on the north of the island is one such flow, where the palaeo-valley erodes into the surface of the western granites. Bloodstone Hill and Fionchra also owe their flatness to lava flows filling eroded valleys. Huge as it had been, the Rum volcano was being eroded away rapidly and had been reduced to only a vestige of its former self by the time 'flood' basalts began to cover Skye and, as we will describe, even far beyond. Rum had had its day.

The real mystery of Rum, though, is not the size of the original volcano, its age or even how much of it was washed away. What is special for geologists, and brings them here from around the world, is the layering of the gabbro. We find that these rocks, originally a liquid magma, were forming only about 2–2.5km below the surface. Oil wells are regularly drilled down to four kilometres, so we should

Figure 4.8
The summit of Askival disappearing into the mist. The original volcano towered perhaps two kilometres above today's remnants.

be able to drill into a magma chamber. The Japanese were going to do just that on Mount Fugendake just above Okinawa, although it will probably take a few years. However, the drilling will not explain why the gabbros on Rum are so intricately and finely layered. Glacial erosion picks out only the coarsest units, as the scramble up to the stepped peaks of Askival and Hallival makes clear. But stop on the way up and look at any weathered rock face (it must be weathered, as fresh surfaces show nothing). The col between the two peaks is a perfect place. On the large exposed weathered surfaces, an intricate layering is picked out, even down to the finest millimetre; the whiter layers stand out, the speckled darker layers are slightly more weathered. The detail is incredible. There are contorted layers, wavy layers, layers tilted and even mixed one into the other. All this layering and structuration continues through the entire vertical thickness of more than 800m of gabbro, delicately indicating the internal activity of the ancient

magma chamber. But the cause of such fine and intricate layering is the real mystery, especially as it is happening inside a chamber of searingly hot magma. Clearly, whatever the cause it has to be common and continuous in a volcano. Modern measurements suggest that cause. Although it is not immediately realized, all the rocks in volcanoes are layered, not just the magma chamber; the structure of the volcano itself is a pile of lava layers. It is therefore logical to link the two, extruded lava layers with internal magma layers. Let us explain.

The one event that is measured on Hawaii on a daily basis is the bulging and collapse of the surface: bulging as the magma temporarily fills the shallow chamber, collapse when lava is erupted. This happens over a cycle of the order of months. Inside the magma chamber this must be having an effect. As the volcano gets bigger, the magma chamber gets higher and grows vertically upwards. It does this by

Figure 4.9
The incredibly detailed layering (originally horizontal) of the Rum gabbro. The layering formed from molten magma inside the magma chamber and is an intimate diary of the eruptions of the 65-million-year-old volcano (courtesy of Lorne Gill, SNH).

cooling from the bottom. Geologists who study the details of the layered gabbros find that the individual layers are made up from crystals with different temperatures of formation and different densities. The darker, speckled layers contain a lot of olivine, a dense green mineral that crystallizes at high temperatures, and also sometimes chromite (chrome silicate), which is very dense (and valuable). The white layers contain amphiboles and feldspar, a sombre-coloured and white mineral respectively, both less dense than olivine and forming at a slightly lower temperature. In other words, in the molten magma there are crystals that form first and have time to settle through the chamber to collect on the bottom, as too much sugar does in a teacup. Layering represents differences and will only form when magma conditions vary, for example, when there is a change in pressure or temperature or a new influx of magma. Measurements on Hawaii, as we have described, suggest that magma is flowing into the shallow chamber constantly but that lava emissions occur only from time to time, on a monthly, yearly or even longer basis. Conditions in the shallow magma chamber vary constantly but will be especially affected by eruptions; it is then that large changes in pressure and temperature will occur, and these will cause the fine layering. The suggestion, therefore, is that each fossilized layer is linked to an eruption, and that even the importance of the event may be indicated by, for example, its thickness or character. Like tree rings, the fine layers of the Rum gabbro contain the intimate 60.5-million-year-old details of the growth of the volcano. The usual explanation links the inflow of new magma from below the volcano to the layering of the gabbro in the magma chamber, not the outflow from the chamber to the surface. So let us use some numbers in support of the link to eruption rather than to inflow. The Rum volcano probably lasted between 250,000 and 1,000,000 years so that in the 800m-thick layered sequence that we can see and measure, 2.5cm of gabbro represents between 8 and 32 years. Magma inflow does not vary on this timescale: outflow does. The events and fine layers are not regular; lava can erupt every month or year or there may even be a gap of centuries. No one has yet tried to interpret this layer history in terms of time, but it will certainly help in understanding the eruption patterns of modern volcanoes. The daily diary of a volcano is an important document that usually takes years of observations to compile. Here it is, intricately fossilized in the Rum rocks like a Rosetta stone, quietly waiting to be deciphered.

Superimposed on the fine-scale layering of the gabbro is a physically much thicker set of variations that represent a much longer timescale and give rise to the benches we first saw in Coire Dubh at the top of the path out of Kinloch. These are the 50m-thick units that are mapped around the island, and there are 16 of them. From the average thickness of the units and the time the volcano lasted, each represents 15,000–60,000 years. So although the Rum volcano is varying on a daily or monthly basis, it is also repeating itself on this much longer timescale. If we can understand this scale of cycle, it may help to predict the long-term activity of volcanoes today. Before being able to look at this repetition scientifically, however, we need to look at the rocks that make up the magma chamber, the gabbros, and a new type of analysis that is used to understand them.

The fine-scale gabbro layers may tell us what happened as the magma chamber emptied but the gabbro itself, its composition and its geochemistry, can tell us about what was happening deep below and what sort of magma was flowing into the chamber. This interests geologists because volcanoes provide a chance of sampling what the Earth is made of. A century ago volcanoes were thought of as safety valves getting rid of the Earth's extra heat, a kind of global sweating system. Now volcanoes are seen as natural leaks from the mantle, providing a perfect system for sampling it. On Hawaii, instruments detect the magma coming from 60km or more below the surface, and a similar depth is proposed for Rum. At this depth, where pressure is very high, rock cannot melt entirely and only a fraction, perhaps as little as 1% but possibly up to 10%, becomes liquid. Once melted, the liquid is less dense than the surrounding solid rock and begins to rise naturally upwards at speeds, it is suggested, of 1–2m per minute. This means it takes as little as 20–40 days, a mere month, to get to the surface from 65km down. The liquid that reaches the surface, therefore, only represents the tiny, melted part of the mantle and not the whole rock. It is a little like distilling whisky, when only the alcohol goes off in the vapour during distillation; lots more, essential for the making but useless for the drinking, stays behind. Volcanoes, if you will, are the whisky of mantle distillation. We can learn more of this process when we use some clever modern analytical methods involving new measurement techniques and complicated, computer-driven, robotic machines.

The targets of these new analyses are trace elements, especially the rare earth elements (REE), of which there are 14, with atomic numbers between 58 and 71. Most of them are generally unheard of, such as neodymium or dysprosium or even europium, but they are detected and measured in modern plasma mass spectrometers. The elements are

Figure 4.10
Enlarged, transparent slice of typical gabbro (thin section). The bright colours are olivine crystals, the white and grey lined crystals are feldspars. Horizontal bar 1mm.

very rare, evidently, and are not abundant enough to form minerals themselves, but importantly, they are grudgingly incorporated into other common mineral crystals in a very specific way. Because the rare earths are not generally incorporated into crystal structures, they are called 'incompatible', the significance of which we will describe. The rare earth elements come in a range of ionic sizes. The elements with smaller ions, non-intuitively called heavy rare earth elements (HREE) like Yb, Lu and Y (ytterbium, lutetium and yttrium) are more easily incorporated into a crystal and are called less incompatible; larger sized ions from the light rare earth elements (LREE), however, like Ce and La (cerium and lanthanum), are progressively more difficult to incorporate as their size increases and are therefore called more incompatible; a kind of chemical natural selection and – dare we say it – 'size matters'. The more incompatible elements are the last to be selected, and the final fluids from a magma to crystallize are rich in the most incompatible elements. Begin to re-melt this rock and everything happens in reverse. The first fluid produced by partial melting is rich in the more incompatible LREE, but if melting continues and more liquid melt forms, the relative amount of these elements decreases. That is, if only 1–2% of a rock melts, then the liquid produced is much richer in incompatible LREE than if 10% melts. A set of typical values is used in order to compare these variations; after all, what is 'rich' and what is 'poor' in terms of rare earth element content? The set of normal values is based on measurements of meteorites, specifically the stony meteorites that are considered to represent the material that originally coagulated to form the Earth. In other words, stony meteorites represent the primitive values that the Earth would have had before it started to separate into layers of crust, mantle, core, and so on. So 'rich' in rare earths means more than in primitive material; 'poor' means less than. Interesting, but what is the significance of all this and how is it applicable to the Rum volcano?

To say that the magma feeding the Rum volcano comes from a depth of 65km is like saying all the cars on the M25 at a certain moment come from Essex. The magma feeding Rum certainly came from below the island but equally certainly it came from a range of depths. Partial melting varies with depth, the shallower intervals producing a higher percentage of liquid melt than deeper intervals, principally as a result of the effects of pressure. In other words, higher pressures (i.e. greater depths) prevent melting. If we bring the rare earth elements into this argument, then magma that has a deep source will come from a smaller volume of partial melting and will be richer in the more incompatible, lighter rare earth elements. A shallower source from a greater volume of partial melting will be less rich in incompatible rare earth elements. As the M25 analogy suggests, there will normally be a lot of mixing, and shallow and deep sources will be mixed together. However, there may well be tendencies. Just as most of the cars on the motorway early in the morning will be local commuters, so the early magmas will tend to be from greater depths and the later from a range of depths. The rare earth elements from the gabbros of Rum are poor in light rare earth elements (those that are more incompatible and the last to be incorporated), which may be because of the depth of origin of the magma or its type. However, what is significant is the variation in these elements through the large, 50m-thick cyclic units (numbered on maps).

At the base of each cycle the LREE (those most incompatible) have higher values than at the top of the unit. At the same time there is a decrease in magnesium upwards, as well as a change in certain strontium isotopes, all of which can be interpreted in terms of depth of origin of the source magma. At the base of each cyclic unit, a deep source is indicated with a small amount of partial melting (high LREE). As time goes on, a shallower source with a greater volume of melting is indicated (more HREE), which then mixes with the deeper-sourced material as it passes through. There appears, therefore, to be a deep-seated pulse to the volcano, although why it should have a beat of 15,000–60,000 years is not known; it means the deep source begins a new melting cycle every 15,000–60,000 years. The reasons for this cycle are to be found deep within the Earth, not near the surface. At the moment, no one seems to be looking in the right place. The Rum volcano, like Pepys, has left us with a fascinating diary of its activity. However, the translation of what is written in the rock diary is not yet agreed upon; the language for the moment is obscure. And of course, the ideas have to be rigorously tested in the accepted scientific way.

The Tertiary of the northwest Highlands should not be remembered just for its volcanoes; fun as they may be, they form only a very small volume of all the volcanic material that was erupted. The greatest volume of the episode, by far, is represented by 'flood' basalts, the basaltic lavas that spread in large volumes over the land. If volcanoes are the single malts of mantle melting, flood basalts are the blends. To look at these we must leave Rum, cross to the Isle of Skye from the Kyle of Lochalsh on the mainland to Kyleakin in Sleat on the island, before travelling on to the north.

The cost of buying the Skye volcano from the Clan MacLeod in 2001 – for it was for sale – was £10 million, an insignificant 17 pence for each year of its 59-million-year age (59Ma). A bargain? Nobody thought so except the Scottish Executive (government), so we in Scotland will all pay a pound or so to John MacLeod, the clan chief and owner. As volcanoes go there must be cheaper, but for character and renown it is probably untouchable, so perhaps it is worth it. Skye is an island of two Jekyll and Hyde halves, south and north: the volcano and the lavas; the visited and not visited; the climbed and the uninteresting; the 'for sale' and the 'not for sale'; the Cuillins and the Quiraing. The contrast is startling. Everyone has heard of the Cuillins. Who, outside of Skye, has heard of the Quiraing? The fishing village of Portree, the capital of the island, is sited almost at the divide between the two halves, between the dramatic hills of the Cuillins, the old volcano, and the flat, peninsula fingers of Duirinish, Waternish and Trotternish, which make up the northern lava fields of

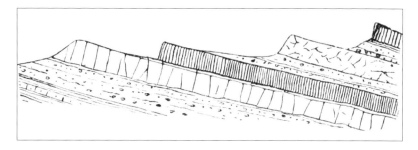

Figure 4.11
Layered lava creating trap scenery (after Geikie, 1909).

flood basalts. The village is built on the lavas but in the cliffs, beaches and bays around the fishing harbour, the bottom of the lava pile can be seen. The rocks are rather uninteresting and the beaches covered in dead sheep, dead seals and plastic debris, mostly from boats. In 2018 there was a groundswell to do something about the plastic on beaches. It will be interesting to see how this progresses. North of Portree the nature of the lavas is expressed in what is known the world over as trap scenery (trap is Dutch for step) – flat, sloping fields separated by shallow, abrupt vertical cliffs, making giant steps out of the land. They are called MacLeod's Tables on Skye, one lava flow stripped away from the other by weathering.

The Tertiary lava pile that remains covering the north of Skye is, even today, over 1000m thick and dips gently westwards, leaving the land with jagged cliffs and black, gritty beaches. But the soil is good and grass and wild flowers grow. The lavas are between 60.5Ma and 59Ma old (precise dates are difficult to measure), although their eruption came after the Rum volcano (60.5Ma) and before the beginnings of the Skye volcano (59Ma), now the Cuillins. The lavas came not from central vents but from linear cracks and fissures, and are spread evenly over the land surface, typical characteristics of flood basalts, similar to those of Iceland today.

Running from Portree northwards through the middle of the Trotternish peninsula, a ribbon of high, east-facing cliffs weaves to the top of the island, crumbling from erosion around the Old Man of Storr, and culminating at the Quiraing in a cliff face 50m high. The evident layers in this cliff are lava flows, irregularly stacked one on top of the other, 2–10m thick. The thin, bright red layers separating many of them are old soils, very rich in iron, which developed between eruptions. From their age and thickness, the Skye lavas seem to have accumulated at a rate of about one metre every 250 years, which means perhaps one flow locally every 500 years or longer – not often – and certainly leaving enough time between eruptions for the bright red soils to develop. The base of the cliff at the Quiraing and at Storr can be taken as the bottom of the main Skye lava pile, since below are mainly older, non-volcanic rocks injected by basalt sills and dykes, feeders for the lava flows above. Exploring the lava pile is disappointing. The feel of primitive fire and hot lava oozing from deep cracks in the earth is gone and all that is left are crumbling, grassy cliff faces in which little can be seen of the original flows. The black cliffs of the Old Man of Storr and the Quiraing are imposing but unscalable. The basalts impress more with their scenery than with their outcrop. But the truth is that they are central to the early Tertiary story. Volcanoes are extrovert and expressive, basalts are boring and introvert but, as is true in humans, the quiet character is often the one with the deepest meaning and the one with most to give. Quite recently, flood basalts such as those on Skye have been discovered over a vast area below the sea to the northwest of Scotland, and all dating from the early Tertiary. Today they are below the sea under one to two kilometres of sediment and can only be detected using sophisticated, modern geophysical methods. Despite this, a fascinating picture is being slowly put together, which shows that Skye and the other islands represent just a few brush strokes in the bottom right-hand corner of a massive canvas.

Because oil was found in the North Sea, exploration companies wondered if there might also be oil on the continental shelf off northwest Scotland. Some 45 years ago they had no idea and began searching. Because this requires some sort of sovereignty over the waters, the tiny volcanic outcrop of Rockall, 500km offshore, which only just breaks the waves, was 'colonized'. To do this, the marines blew the top off the rock to make a flat bit and then, to demonstrate that Rockall was a British colony, bolted a navigation light to it: a bright colonial light. Greenpeace ignored the light and invaded, attempting to assume the sovereignty for themselves and so control the exploration. It would have been Greenpeaceland, since the name Greenland is already taken (used very effectively by Eirik the Red in 985AD as a white lie to entice Norwegians to colonize it). Greenpeace fell off the rock after a while but considered they had made their point (which was? And who makes their policies anyway?). Regardless, the seas and the shelf began to be explored using seismic boats and some airborne equipment. Seismic boats use sound to penetrate below the Earth's surface and obtain an image. The boat tows behind it a sound source and a lot of very sensitive microphones, or hydrophones (because they work in water). The sound comes from a gun that blows bubbles of compressed air under water, the collapsing of the bubble giving the right kind of sound pulse to go down to the sea floor and into the various rock formations below. It is then reflected back to the surface to be detected by the hydrophones. With very sophisticated, computerized processing of the hydrophone records, a long vertical sound image, or seismic section, of the top 7–10km of the Earth's rocks can be created. The hydrophone equipment is so sensitive that it can easily detect whales as they sing to each other, which it does, as they do. The different rock layers below the sea floor can be seen on the seismic sections and their shape and character plotted on maps. Encouraged by the seismic records, some years ago the oil companies began to drill the rocks and oil was found. This stimulated more exploration, more seismic, and more drilling. In the year 2000 the targets were below the seas around the Faroe Islands, which of course 'belong' to the 20,000 Faroese. Like the north of Skye, the Faroes are made from a pile of Tertiary lavas, even though they are 500km away to the northwest. As on Skye, there are older, non-volcanic rocks below the lavas, which is what interests the oil companies, as this is where the oil might be found. Exploration drilling in the seas off the Faroes is continuing slowly and may provide the islanders with a different income to fish. Rockall, meanwhile, remains a republic.

In seismic records, lavas are usually easy to recognize; they are hard and dense and are a good reflector of sound. This means that they leave a 'bright' line on seismic sections. In addition, lavas are often drilled through in oil exploration wells. Pieced together, the geophysical and the well information show that the lava fields are immense. They stretch from 70°N, the latitude of northern Norway, to 57°N, around the latitude of Skye, over 2500km, in a broad band that was originally 500–700km wide but which is now split open by the North Atlantic. The western side runs down the east coast of Greenland and the eastern side runs from the west coast of northern Norway to the Faroe Islands and Rockall. The area is as big as the whole of the British Isles and it is as if all of Britain were covered in lava 450m thick, a volume of 6.6 million km^3. Such huge volumes of lava on what was originally land are unusual anywhere on the Earth's surface, and an explanation has not yet been entirely agreed upon. The lava fields of Skye and Mull and the volcanoes of the islands are indeed just an edge to this vast offshore event. But they are nevertheless part of it, as the similarities show. In the shallow seas between northwest Scotland and the Faroes, besides lava, there are at least 20 more volcanoes similar to those on the islands. This is a seismically partially surveyed area and there are obviously still a lot more to be found. One famous volcano in this area, named Erlend, was discovered a little more than 45 years ago from airborne gravity and magnetic readings. It is quite like Rum in that it has a very dense, very magnetic rock at its core, interpreted as gabbro. This leaves a signature high gravity and high magnetic anomaly on surveys. It was these unusual readings that led to Erlend's discovery. A seismic survey of the volcano made later showed that it stood at least 1500m high, despite the top having

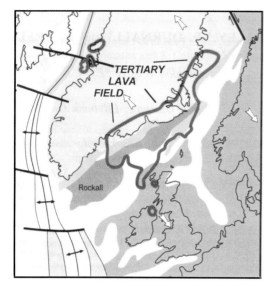

Figure 4.12
The huge extent of the Tertiary lava field illustrated on a plate reconstruction at 55Ma – just before the North Atlantic opened. The small red blob to the northwest of Scotland is the on-land outcrop area that includes Skye and Rum. White arrows indicate subsequent relative plate movement (modified from Naylor et al., 1999, courtesy of the Geological Society London).

been eroded away, and had a base more than 20km in diameter. The vent is quite clear on the seismic sections, more than a mile wide and 400m deep. Based on an interpretation of the seismic survey, the flank of Erlend was drilled in 1979 by a company hoping to find oil, but instead, and to their regret, they found only a volcano and a lot of lava.

The huge flood basalt lava fields were a signal for the creation of the North Atlantic Ocean. At the time when the Rum, Skye and other island volcanoes were active in the late Palaeocene, 60–56Ma ago, the European continent was connected directly to Greenland, and the northwest of Scotland was only 400km or so away from the Greenland coast; there was no North Atlantic Ocean (Figure 4.12).

The distance is now 1600km, meaning that the two halves have been separating at a speed of 30km per million years (approximately 2.5cm/year). Samples of lava that have been examined from offshore oil exploration wells, from the Faroe Islands and from east Greenland, as well as from northwest Scotland, all give ages older than the opening of the Atlantic, that is they are older than 55Ma, the moment when the splitting actually took place. Volcanoes were erupting and flood basalts were flowing over the surface of Scotland, Greenland and offshore Norway for nearly 10 million years before the ocean eventually broke open: but why? The huge volume of lava erupted is very unusual and requires a special explanation. The idea, but by no means accepted by all, is that an unusually hot area of the crust, a 'hot spot', used to lie to the northwest of Scotland. Hot spots are caused by 'mantle plumes', which are upward flows of extremely hot material that come directly from very deep in the Earth, perhaps 500–600km down, and is why they are so hot. The material from mantle plumes is supposed to move vertically upwards, unusually quickly and, moreover, to stay geographically in one place. That is, the 'hot spot' stays still while continents drift across it, just as a saucepan can be moved across a gas burner.

The mantle plume that caused the 'hot spot' below the northwest of Scotland, is now under Iceland, although it is Scotland that has moved (southwest), not the 'hot spot'. A mantle plume causes volcanism because it raises the temperature underneath the crust, causes lots of melting and, as explained for Rum, once a rock melts it will rise naturally to the surface, like the bubbles in a lava lamp. This rising Tertiary plume only has a diameter of 100–200km at depth, but below the crust it spreads out like a mushroom cap and creates a 'hot spot' with a diameter of over 1000km, causing the many and widespread volcanoes, but especially causing the huge outpourings of flood basalts. What aids the eruption of the lavas is that, as in anything boiling, where a hot current rises, the surface rises too. Over the mantle plume, the Earth's surface rises like a boil, becomes stretched, cracks, and out pour the lavas. The stretch marks, in the form of dykes, that is thin, linear, vertical intrusions, can still be seen today, and of course the lava fields are evidence of the outpourings – a fine story. There are many 'hot spots' around the globe; even Hawaii is on top of one, so they certainly do produce lavas, but they are not necessarily the reason for a continent to break up. As present ideas go, Rum, Skye, Mull, Ardnamurchan and the rest are all products of a late Palaeocene 'hot spot' and associated mantle plume. Today we can climb the Cuillins, and visit the island of Rum, as a result of events that occurred deep inside the Earth, 60 million years ago. Although we now have an explanation for our little bit of volcanic Scotland, there is still no explanation for the North Atlantic Ocean; no reason for the continent to have split apart. When it did, however, all the eruptions in Scotland stopped.

The recent discovery (2017) of a thin layer of meteorite impact ejecta very close to the pile of lava flows, at two sites on the Isle of Skye, has added to our storyline another possible cause for the initiation

Figure 4.13
A seismic section from northwest Scotland running NW–SE for 160km. The irregular red line is the top surface of the Tertiary volcanic lavas – interval 4. The explosive ash deposit (tuff) indicating the initial split of the North Atlantic Ocean is the layer above this horizon – interval 5 (after Tate et al., 1999, courtesy of the Geological Society of London).

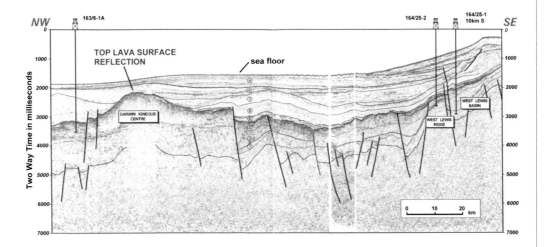

of volcanic activity. It is only the second occurrence of meteorite ejecta discovered in Scotland, the other being in the Torridonian and discussed in Chapter 1. This time, as well as Reidite, there are unmelted crystals from the meteorite itself. The ejecta do appear to contain some pieces of slightly earlier lava flows, suggesting that the eruptions may already have just started when the meteorite hit. It does pose questions, however, as to whether the impact contributed to the outflowing of the lava and where the impact crater may have been. Once again more research is needed, and there is more to learn.

Fifty-five million years ago the continent broke open and a series of great explosive volcanic events spread ash and volcanic dust around the globe. Over the immediate area around the split we still find the ash and dust, now fossilized into rock – tuff, in scientific terms. It can even be found (as rock) on the shores of southeastern England, in northern Denmark and, more significantly, in all deep oil exploration wells drilled in the middle of the North Sea, off the coast of Norway, or around the Faroe Islands. The layers of ash and dust from the explosions, tuff layers, are easily recognized, even though they are now buried under 1000m or more of younger sediment. When the ash dust fell onto the sea surface following an explosion 55Ma ago, it simply dropped to the sea floor, the larger, heavier particles falling first. Today we find thin, fossilized layers of graded tuff beds, that is, beds with coarser grains at the bottom and finer grains at the top, 3cm thick, each representing one of the huge explosions from the opening ocean. Danish geologists have even numbered the layers, or counted the explosions if you like. There are nearly 200 onshore in Denmark, but offshore, in the middle of the North Sea, there are at least 500 and together they make up a 50m thickness of debris and sediment: nature's bookmark in the rocks for the opening of the North Atlantic Ocean.

The dust and ash explosions were quenched as the active volcanoes subsided beneath the deepening ocean. The split continued to grow but under water, eventually becoming the mid-Atlantic Ridge as Greenland drifted to the northwest, and Scotland to the east-southeast. As the continents moved, the 'hot spot' stayed where it was and is now beneath Iceland, meaning that the island is still where the line of the split originally started. Today that line passes through Thingvellir, the site of the first parliament in Europe, speakers standing on the ramparts of the faulted lava while listeners crowded on the grassy flats below, each fault a witness of the active splitting. It is quite amazing to think how appropriate this place is for a parliament, a natural split separating two opposing political sides. We now know the precise way that the ocean crust fractures and heals along the split because of the magnetic record left in the lavas, a major discovery only made in the 1960s. Basalt lava contains many iron particles that, when the lava cools, take on the direction of the Earth's magnetic field, a feature which, once frozen, is preserved. Mysteriously and unpredictably, the Earth's magnetic field suddenly changes direction (i.e. polarity), north

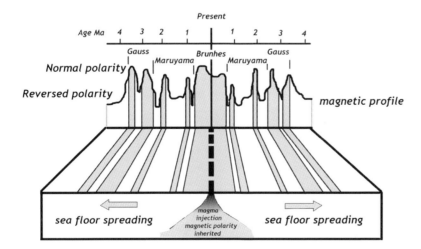

Figure 4.15
The origin of magnetic stripes that allow the ocean floor to be accurately dated. The names are those of the Magnetic Chrons from the last 5Ma, Pliocene and Pleistocene. Brunhes, the present normal polarity chron, began 0.73Ma ago.

becoming south every few million years. When the polarity changes, newly erupted basalt takes on the new polarity, but previously erupted, older basalt preserves the polarity it inherited when it cooled. This means that as an ocean splits and spreads, lines of basalts are left behind with different magnetic polarities, a symmetrical set of lines either side of the split. The past periods of normal and reversed magnetic polarity are recognized and numbered as normal and reversed couplets, each couplet geologically called a Chron. The first and oldest magnetic stripe on the North Atlantic Ocean floor is Chron 24r (r for reversed), which formed between 53.5Ma and 55Ma. One side is found along the Greenland coast and the other along the north Norwegian coast, but it continues just to the west of the Faroe Islands where the ocean floor begins. Present day Iceland contains the sequence of magnetic stripes from Chron 7, nearly 30Ma old, which occurs along the northwest coast and the southeast coast respectively, up to Chron 1n, the present day (by convention normally polarized), which crosses the centre of the island and on which Thingvellir is sited.

The activity deep in the Earth that created the Rum volcano 60Ma ago, that caused the Skye lavas 59Ma ago and the tuff layers 55Ma ago, has moved to Iceland and the Mid-Atlantic Ridge. Today Scotland, like the rest of Europe, continues to move slowly southeastwards. There is an occasional creak or small earthquake, but the active volcanoes are long gone. The beautiful brooding hills of the Cuillins and the rotting cliffs of the Quiraing are all that is left of the mantle plume: the sulphur, smoke, geysers and glaciers of Iceland are where it continues. The Manx shearwater fledglings can rest content in their burrows high on Hallival; the Rum volcano is just a big, cool, friendly fossil.

Further Reading

Books, Pamphlets

Decker, R. & Decker, B. 1979. *Volcanoes.* New York: W H Freeman & Co. pp.321. ISBN 0 7167 2440 5.

Emeleus, C.H. 1997. *Geology of Rum and the adjacent islands.* Memoir of the British Geological Survey. Sheet 60 (Scotland). ISBN 0 11 884517 9

Emeleus, C. H. and Bell, B. R. 2005. *British Regional Geology: The Palaeogene Volcanic Districts of Scotland.* 4th Edition. British Geological Survey, Nottingham. pp.213. ISBN 0 85272 5191

Emeleus, C.H. & Troll, V.R. 2008. *A Geological Guide to Rum: The Paleocene igneous rocks of the Isle of Rum, Inner Hebrides.* Edinburgh Geological Society in association with NMS Enterprises Ltd.

Goodenough, K & Bradwell, T. 2004. *Rum and the small Isles.* Scottish Natural Heritage. pp.39. ISBN 1 85397 370

Holmes, A. 1993. In: Duff, D. (Ed.) *Principles of Physical Geology.* London: Chapman & Hall. 4th Edition, pp.791. ISBN 0 412 40320

Juteau, T. and Maury, R. 1999. *The Oceanic Crust, from Accretion to Mantle Recycling.* Chichester, UK: Praxis Publishing. (Translation). pp.390. ISBN 1 85233 116 X

Magnusson, M. 1997. *Rum: Nature's Island.* Edinburgh: Luath Press. pp.137. ISBN 0 946487 32 4

Sigurdsson, H. 1999. *Melting the Earth.* Oxford: Oxford University Press. pp.260. ISBN0 19 510665 2

Stoker, M.S., Hitchen, K. and Graham, C.C. 1993. *United Kingdom offshore regional report: the geology of the Hebrides and West Shetland shelves and adjacent deep-water areas.* London: HMSO for the British Geological Survey. pp.149. ISBN 0 11 884499 7

Further Reading (continued from page 135)

Scientific Papers

Drake, S., Beard, A., Jones, A., Brown, D., Fortes, A.D., Millar, I., Carter, A., Baca,. Downes, H. 2017. Discovery of a meteoritic ejecta layer containing unmelted impactor fragments at the base of Paleocene lavas, Isle of Skye, Scotland. *Geology* 46 (2).

Naylor, P.H., Bell, B.R., Jolley, D.W., Durnal, l. P. and Fredsted, R. 1999. Palaeogene magmatism in the Faeroe-Shetland Basin: influences on uplift history and sedimentation. In: Fleet, A.J. & Boldy, A.R. (Eds.) *Petroleum Geology of Northwest Europe. Proceedings of the 5th Conference.* pp.545–558. Geological Society of London.

Tate, M.P., Dodd, C.D. and Grant, N.T. 1999. The Northeast Rockall Basin and its significance in the evolution of the Rockall-Faeroes/East Greenland rift system. In: Fleet, A.J. & Boldy, A.R. (Eds.) *Petroleum Geology of Northwest Europe. Proceedings of the 5th Conference.* pp.391–406. Geological Society of London.

Chapter 5   THE COMING ICE AGE

Past and future climates in the Highlands

Winter and summer, spring and autumn come so regularly that we overlook their fragility, the delicate balance that climate is, being quite unaware of what even the slightest changes will do. If the average summer temperature in the Highlands were to drop by 4.0°C, the difference between a tee-shirt and a sweater, all the Highland Munros would be permanently snow-covered, glaciers flowing from their flanks. Put another way, the height at which permanent ice would form in the Highlands today is only c.250–300m above present highest peaks. Even now, in some years small amounts of snow last well into summer in gullies on Ben Nevis. The balance is very delicate. In the past, of course, such calculations were unnecessary, and during the depths of the last ice age 25,000 years ago, a huge ice-sheet covered all of Scotland and even northern England down as far as York. This idea is a familiar one today, though only 160 years ago such a proposition was unknown. The first person to propose an ice age, even for Scotland, was Louis Agassiz (1807–1873), a Swiss natural scientist who became convinced in the 1830s, while working on glaciers in his home country, that ice had once covered large parts of Northern Europe. He visited the Highlands in the autumn of 1840 with William Buckland (1784–1856), then President of the London Geological Society. *The Scotsman* newspaper published a letter from Agassiz in October of that year, written from Fort Augustus:

Figure 5.1
More often than not in the recent past, the Scottish Highlands have been covered in permanent ice, glaciers occupying the valleys.

> ...*at the foot of Ben Nevis, and in the principal valleys, I discovered the most distinct moraines and polished rocky surfaces, just as in the valleys of the Swiss Alps in the region of existing glaciers, so that the existence of glaciers in Scotland at early periods can no longer be doubted.*

Interesting that a major, new scientific discovery was announced in a letter to a newspaper: unthinkable today, of course. Being the first time that the glaciation of the Highlands was scientifically proposed, there were doubters and many were not convinced, still insisting on 'a great marine submergence' to explain the Highland scenery. In fact, the ice age theory was not entirely accepted for another 20 or more years. In Scotland, this eventual acceptance had a lot to do with Archibald Geikie (1835–1924, the same Geikie as in Chapter 2) who in 1855, when he was just 20, had been one of the first geologists to be appointed to the newly created Scottish Geological Survey. Although Geikie was at first not convinced of Agassiz's glacial theory, what he saw in the mountains during his geological work for the government altered his views and he became a zealous convert. With his brother James (1839–1915), from the mid 1860s he set about describing the glaciation of the Highlands and effectively persuaded most scientists of its reality. What immediately impressed these geologists was, of course, the fact that great ice-sheets had existed in the past where now there were none. But it was especially the huge feats of ice erosion that marked their imagination. The gouging out of great deep, U-shaped valleys, the grinding of rocks to leave scratched and polished rock surfaces, huge, stranded boulders and the massive amounts of dumped rock debris; this is what impressed them. This is what they worried about. Not now. What impresses us now, what we worry about now, is the cause of it all: the change in the climate.

The great extent of ice-sheets from Russia to North America in the ice ages, the landscapes of ice erosion in the Alps and the Rockies and the heaps of dumped sediments in fjords and bays are so familiar to us today that we think no more about them. Now we have become fascinated by, and rather nervous about,

Figure 5.2
What impressed the early workers was the power of ice to erode and to transport great boulders. What worries us now is climate change and our influence on it (after James Geikie, 1894).

climate—and of course whether we ourselves are changing it; we are fascinated by global warming. But is it really happening, and what can the Highlands tell us about climate change? This chapter is about both of these things: climate change and what can be seen of it in the Highlands, especially of the ice ages.

Plagiarism by writers is common: everything has been written before. So the title to this chapter 'The Coming Ice Age' is stolen from

141

a book written in 1896 by a generally unknown American sailor and scientist, CAM Taber. For him, writing 100 years or so ago, the next ice age was imminent. Now, with global warming, it is ice melt that is imminent. Science has fashions. Now fixed continents, now continents moving everywhere: now ice age, now global warming. History tells us that just because an idea is fashionable does not make it right – perhaps even the opposite. Fashionable ideas are likely to be wrong; science does not advance by democratic vote or public opinion poll. That is for politics. But global warming has an even greater handicap than being fashionable; it now has a huge vested interest from environmentalists, scientists, politicians, software writers, technical and engineering companies, insurance brokers and even bankers. Global warming is business. It is not possible to debate the subject any longer without substantial overtones.

The Highlands have their part to play in this debate. We use windmills and small-scale hydro schemes to try to defer the carbon release of fossil fuels, and the peat bogs hold on to a huge carbon reserve. Does any of this help? We can but do our best to monitor developments. The number of books on the subject of climate change is huge, but they are all the same: global warming the facts; global warming the human reasons; global warming the necessary remedies; and especially, global warming if we don't do what we are told. But are we giving ourselves a planetary importance that we do not have, first of all to be able to affect global climate for the 'worse' and then, when we have so decided, to affect it for the 'better'? (Note that better and worse are not scientific terms; they reflect what is supposedly better or worse for people). From a geological perspective, this is, of course, laughable. From a human perspective this is replacing the previous God of Noah's great flood with a new god, *Homo sapiens*, the species: both equally silly prospects. King Canute will always get his feet wet.

One more thing. As science progresses, more and more specialists and specialist subjects are invented. It is a source of wonder to today's scientist that Leonardo da Vinci was not only a supreme artist and sculptor, but also architect, engineer and inventor of war machines. Art and science today are very far apart. Specialization in geology, as in other sciences, has had the effect of creating sub-cultures all over the subject. Like street gangs, each specialist group has its own language which has to be understood to become a member. The Geological Society of London has 22 such groups, and in every university this is repeated: the palaeontology gang at this end of the corridor, the geomorphology gang at the other and the geophysicists in the middle. This is serious. Just as street gangs hinder proper social functions, so the 'specialists' hinder the proper functions of science. And this is even more obvious when it comes to major subjects: cross the divide between geology and archaeology or astronomy or climatology, and you will be treated as a pitiful amateur. You don't have the right terms or the same language and you don't know the right people. What has this got to do with ice ages and global warming? Everything, because the study of ice ages sits perfectly between geology, archaeology, history, climatology, even astronomy, as well as human anthropology. Humans have lived and developed through the ice ages so that we have left a record, first artefacts then written accounts and now instrumental records. Humans were affected by the ice age climate and now, so it goes, the climate is being affected by humans. Here, then, is the problem: geology runs into archaeology, which runs into history, which runs into the future; the human record is in a continuum with the rock record; a whole rainbow of specialists is involved. So let us settle into an unaffected look at the ice-world of the Highlands with a peek at the past, a demonstration of the present and a stab at the future. Rocks will be the past, humans the present, instruments and software the future. There will be no regard for subject limits, big business, environmentalists, politicians or any other self-interest group.

Since the great work of Archibald and James Geikie in the 1860s–1870s on the glacial history of Scotland, much has been done and learnt, but as we shall see, the land record that has been used to recreate past ice-worlds is very broken, punctuated and hugely incomplete, even in the Highlands. It is only in just the last few years that revolutionary information has become available from beyond Scotland, which gives not only a complete record of the ice ages, but a year-by-year commentary on ice age climate. A glacial weather forecast in reverse, as it were. The information comes from two principal sources: cores of ice from Greenland and Antarctica and cores of sediments from the deep oceans. The interpretation of these records is elegant science. We now have a complete ice age diary that we can annotate with the occasional events from the land record of the Highlands. Let us first look at this new information.

Progressively, over the last 250,000 years, nearly three kilometres of ice have accumulated to make up the Greenland ice cap. Every year new snowfall collects and, under its own weight, becomes gradually compressed into fine layers of solid ice, each one a perfect,

Figure 5.3
A fossil climatic diary. Annual layers in the Greenland Ice Sheet Project core (GISP). Red lines indicate light summer layers; the winter accumulations are dark (courtesy of NOAA, photo Anthony Gow).

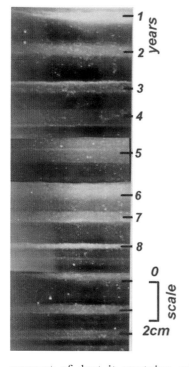

frozen record of a year of climate: the temperature, the gases in the atmosphere, the wind-blown dust and the amount of snow. This has now been continuously cored, notably at the Greenland Ice Core Project (GRIP), finding nearly two miles of ice layers. Count back from the surface and you have a yearly record, accurate to 1% as far back as 10,000 years, slightly less accurate back to 100,000 years but present up to 250,000 years, and all with so much information. For instance, the thickness of each layer is an indication of the amount of snow that fell that year. The colour of the ice, measurable by laser, is a reflection of the amount of dust it contains, which may have come from known historical volcanic explosions. It is even possible to analyse the past atmosphere. When snow falls it is quite loose and has many holes filled with air. Gradually as the snow compacts, the air is expelled, but at around 40m depth, some is permanently enclosed in the ice as small bubbles. These bubbles still contain the original atmospheric gases, greenhouse gases if you will, such as carbon dioxide and methane, which can be extracted and analysed: a palaeoclimatologist's dream.

But perhaps the most useful analyses come from the ice itself,

Figure 5.4
Fossilized atmosphere, bubbles of air enclosed in ice.

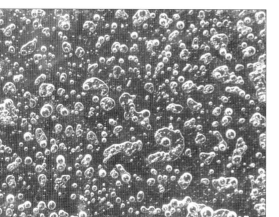

for from it oxygen isotopes can be measured, which can be used as a proxy for past temperatures. Oxygen exists in two principal isotopes (atoms of the same element with different atomic weights and subtle differences in behaviour, a property exploited by palaeoclimatologists), mainly ^{16}O, but 0.2% exists in the form of the heavier ^{18}O. When seawater evaporates, more water with the lighter ^{16}O goes into the air as it is more easily vaporized, which means the seawater left behind has more of the heavier ^{18}O. Consequently, water vapour in the atmosphere has less ^{18}O than seawater. During condensation, when snow forms in clouds, the isotopes again behave differently, more of the heavier ^{18}O going into the snow since it has a lower vapour pressure. This snow then becomes the ice that is stored. The eventual isotopic differences in the stored ice depend on the temperatures at which both evaporation and condensation take place. The lower the average temperatures, the lower the amount of ^{18}O, by 0.7 of a part per thousand for each degree (centigrade) of decrease: an amazing and accurate thermometer for the past. It is usually plotted as $d^{18}O$, the difference (delta, d) between a reading and the present day standard mean ocean water (SMOW) temperature. In addition, although the ratio of $^{18}O/^{16}O$ in rain and snow may vary according to atmospheric temperatures, any differences are evened out when it rains and the water is returned to the oceans. During ice ages, however, there is no return process and huge amounts of snow become frozen in glaciers and ice-sheets, trapping an ice enriched in ^{16}O. The result of this is that seawater becomes progressively richer in ^{18}O; the more ice, the more ^{18}O in the oceans. Which now leads us to the oceanic record.

If the amount of the heavier ^{18}O left in seawater is measured, we will know how much of the lighter ^{16}O has been taken away and trapped in ice, in other words the volume of ice. The ^{18}O in seawater changes by approximately 1 part per 1000 between maximum ice extent and the warmer times when the ice is mostly melted. To estimate past ice volumes, of course, we need to know the ^{18}O seawater values from seawater of the same past times. So how can these be measured, because incredibly they can? When animals living in the sea make a skeleton, they use the elements, including oxygen, from the surrounding water. The calcium carbonate, $CaCO_3$, of their shells, especially of foraminifera, tiny, sea-surface living plankton, will contain the same amount of ^{18}O as the seawater of the time. This means that as the ice volume grows during ice ages and ^{16}O becomes progressively locked up in ice, foraminifera shells will show a progressive enrichment in ^{18}O. The final amount of each isotope will also depend on seawater temperature as well, but this effect will be small. At death, the tiny foraminifera shells fall to the ocean floor. Many are dissolved on the way, but about 1% reach the bottom and are preserved in the thin layers of covering sediment. At water depths of 3000m or more, these sediments stay quite undisturbed for millions of years, collecting slowly shell by shell, layer by layer, a fraction of an inch every millennium. It is this that is cored and analysed. With such a detailed record of d^{18}O in ancient seawaters, we now have an idea of past ice volumes that represent water not returned to the oceans but kept stored as ice on land. Such accumulations, of course, mean that sea-level will drop. We can deduce, as an example, that the eustatic drop in sea-level (i.e. global sea-level change due to glacial advance) at the last glacial maximum 25,000 years ago was 120m. Now we can combine the records from the deep-sea sediments of past ice volumes, sea temperatures and sea-level with the records from the ice cores, providing air temperature and atmospheric gases. The picture that emerges is beyond imagination.

There are three principal ice core records at present, the European GRIP, and US Greenland Ice Sheet Project (GISP) from the summit of the Greenland ice cap and the Russian Vostok core from Antarctica. More are underway. The deep sea records come from the continuously active Ocean Drilling Program (ODP), internationally funded and carried out by the drillship *JOIDES Resolution*. There are many cores from hundreds of very deep ocean sites around the world and the data are publicly available on the internet. All this information will make up our ice age diary.

Figure 5.5
The amazingly detailed record of 250,000 years of climate from the 3km long ice core of the Greenland Ice Core Project (GRIP). The abrupt changes of the Younger Dryas Stadial are clearly marked (modified from Dansgaard et al.,1993). (Ka = 1000 years, BP = before present, H = Holocene)

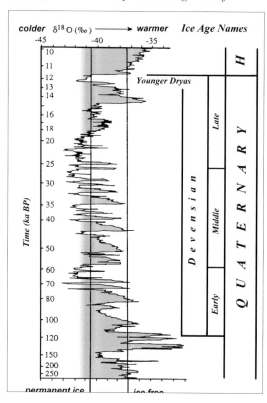

When we look at the oxygen isotope thermometer of the GRIP ice core, the last ice age in the Highlands, called the Devensian, from 116,000 years before present (BP), becomes amazingly real. With a temperature of 3°C less than now, the Highlands of the past were just ice-free. The present d^{18}O level is -35‰, and a 3°C drop means a d^{18}O level of -37‰, which can be used as the 'just ice-free line' (on the Highland temperature plot). A further drop of 5°C, a total of 8°C from today, that is a d^{18}O level of -40.5‰, and ice fields would have been established: a 'permanent ice line'. On the plot of the GRIP d^{18}O record, these Highland values show that there was permanent ice from about 75,000 years BP to 18,000 years BP, then no ice: but ice again from

12,500 to 11,500 years BP. For the last 11,500 years the Highlands and the rest of Scotland have been ice-free. But this is a gross simplification. The continuous trace of the $d^{18}O$ record shows many changes in temperature within the permanent ice band, which were certainly associated with both ice advance and ice retreat every few thousand years. The climate appears to have been very unstable. If this is true, then what is left of all this activity in the northwest Highlands?

Permanent ice leaves traces of erosion; ice melt certainly leaves traces of dumped rock debris. In reality, ice erosion effects tend to accumulate, one glacier taking over where the previous one left off. Differentiating between one advance and the next, or between one separate erosion and another, is almost impossible. With ice retreats it is different. The sediment rubbish dumped by a glacier as it melts will remain, unless the next glacier advances over it, in which case it will be moved on. It is the last glacial melting that will leave the obvious deposits. The reality of this is seen in the valley glaciers around the margins of today's ice-sheets, which are all retreating. When you fly over the glaciers that form fingers along the edge of the Greenland ice sheet, for example, you see heaps of rough stones strewn across each valley, a so-called terminal moraine dumped there by the glacier at its last maximum advance, while the ice is now several kilometres away up the valley. This moraine will remain until the next cold period, when the glacier will return, pick up what it had previously left and push it further down the valley like a brush sweeping the floor. When the ice finally disappears, only the sediment dumped by the farthest advance has a chance of being preserved. This is what has happened in the northwest Highlands. Some of these furthest advances are now under the sea, as in Loch Broom. These are pristine and do not suffer degradation by vegetation and weathering as they do on land, but just have a slight marine sediment cover.

From what has been said, to explore the ice age Highlands we will have to work backwards and look at the most recent glaciation first. The last major ice age in Scotland was the Devensian, when huge ice-sheets developed between 116,000 BP and 15,000 BP (Figure 5.5). However, the last period of cold indicated by the GRIP oxygen isotopes is a short, sharp period between 12,500 and 11,500 BP and is quite remarkable. It is recognized as a 'stade', a very cold period that saw significant ice-sheet and glacier development in the Highlands, although of much smaller extent than in the preceding main Devensian. In Scotland it is generally

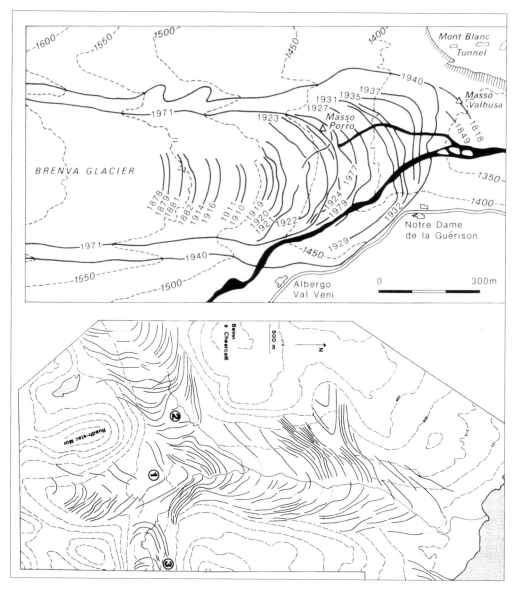

Figure 5.6
Terminal moraine
patterns around
retreating glaciers
today in the Italian
Alps (above, after
Grove, 1988) and
during the Younger
Dryas in Loch Maree
(below, after Bennett
and Boulton, 1993).
Figures courtesy of
Cambridge University
Press.

called the Loch Lomond Readvance or Loch Lomond Stadial. It is
equally recognized throughout Europe and is given the name there
of the Younger Dryas Stadial, Dryas being a small, cold-climate
plant *Dryas octopetala* that became widespread at the time –
'younger' because it had happened before. Using the European
name is best. Although precise dates for the Younger Dryas vary
depending on the type of data used, the event is always clearly
marked – in the Greenland ice, in deep sea sediments, in Swiss and
Scottish lake sediments, tree species in Denmark and, famously, in the
past distribution of beetles across Britain. Some beetle species only

survive within very narrow temperature limits (something to do with their sexual activity – quite understandable) so that their preservation in, for example, lake sediments, can be used to give values of past temperatures, especially those of the summer months when the beetles were sexually active. These 'beetle temperatures', taken from lake sediment cores, clearly identify the cold Younger Dryas Stadial, although not everyone agrees.

Moreover, the beetle studies were made in England and SW Scotland, and do not accommodate the strong thermal gradient from north to south and from lowland to highland Scotland

sniffs one Scottish Highland author. A clear case of scientific nationalism. It appears that midges are also temperature-sensitive and they too are now being studied for past climate changes, which will definitely suit the Highlands and solve the English beetle problem! The undoubted record of the Greenland ice, however, gives remarkable detail to these changes. There is a general decrease in temperature from a high at approximately 14,500 BP with a marked additional drop from 12,500–11,500 BP, taken as the Younger Dryas Stadial itself. There is a difficulty with these relatively precise dates, however. 12,500–11,500 BP is equivalent to 10,500–9500 BC. These dates are sufficiently recent to be used by geologists who use BP, and archaeologists who use radiocarbon dates (which need correcting), and historians who use chronicled years. For the Younger Dryas, historical years are preferred even though they are not actually chronicled. So

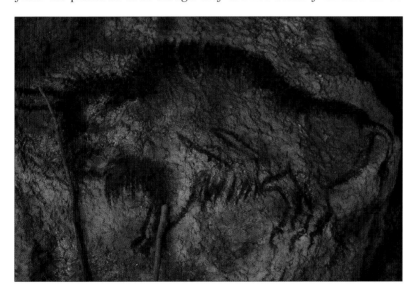

Figure 5.7
Cave painting ceased at the end of the Younger Dryas and civilization as we know it began in present day Iraq. (photo © Alexander Sitnikov, Shutterstock)

we will use the dates from the GRIP ice core of 10,500–9500 BC, although scientifically this is not allowed because these are ice-based years (i.e. BP) and not historical years. However, to be able to relate to human events, using BC/AD is easier.

The Younger Dryas (Loch Lomond Readvance) took place when villages were already being developed in the Fertile Crescent of the eastern Mediterranean (or present-day Iraq). In the ice cores of Greenland, the temperature shows a drop of approximately 5°C during the 2000 years before the actual stadial, from 12,500–10,500 BC and then a further 2°C during the defined stadial (10,500–9500 BC): rapid enough changes. But what is stunning is that around 9500 BC, during just 50 years, temperatures rose a massive 7°C, and this was a global event. The world went from ice-house to greenhouse in a generation! Scientific literature is not supposed to express emotion, so in one of our reference books it simply says:

> *The reason for this event has generated much speculation,...*
> [Like hell! this is a stupendous event, the like of which we cannot imagine. The book calmly continues] *...but as warming occurred coevally in the southern hemisphere it is apparent that it impacted the entire world synchronously and may not have originated simply in the north Atlantic region.*

Environmentalists worry about changes of 0.5°C in 100 years. This is 7°C in 50 years! And the book says primly, 'there is still no explanation'. Experts cannot imagine what could cause such rapid, catastrophic changes on a global scale; it tends to make them nervous. It certainly wasn't caused by cars or consumers, although humans were very much in existence at the time. A critical climatic threshold was clearly reached, but of what? Imagine what this would mean in a community. The past would be no reference to what was to come; one year would change on the previous one and the recollections and experiences of the old people would stand for nothing. Insects would change, animals would migrate, plants would either die or thrive; there would be a massive biological turnover. The emotional effect on humans would have been dramatic and very unsettling. Note that at the same time as this massive climatic event, Neanderthal Man died out, cave art ceased and agriculture began in the Euphrates valley. And what of the Highlands?

During the Younger Dryas Stadial there was an ice sheet over the entire southwest Highlands from Loch Lomond to Loch Maree, over

200km long and 50km wide, small compared to what had occurred in the main Devensian, but huge even so. Separate small ice caps covered the island of Mull and the Cuillins on Skye. Further north, glaciers occupied corries and valleys in the higher areas such as Torridon and the flanks of Conival and Ben More in Assynt. They were there for 1000 years. There are no visible traces of the erosion, slow grinding-down and ice shattering that must have occurred. What we see today is what was hurriedly left behind. The melting of one thousand years of snowfall and ice build-up in just 50 years would have created unimaginable flooding and chaos all across northern Scotland. River flow would have been enormous, vast ice-dammed lakes would have formed and suddenly emptied, great waves of water surging down the narrow glacial valleys. Rock and sediment ground up within the ice would have been dramatically flushed away in flash-flood torrents or dumped in great mounds of waste. The image of a Highland stream, the picture of the water to be used for distilling whisky, is a tiny trickle tumbling between great boulders, the water squeezing through the massive debris left behind by those great Younger Dryas floods. We are always taught that this was because of what the glaciers had left. But look in any glen or strath. The Highland stream, like a Hollywood star, is quite out of place in present reality. What is left scattered across the Highlands is the 11,500-year-old debris of 50 years of catastrophic glacial meltwater flood. *The* great flood.

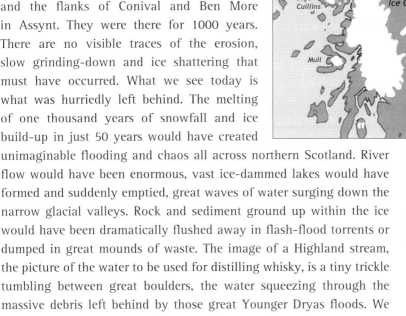

Figure 5.8
The Younger Dryas Ice Sheet (Loch Lomond Stadial) lasted from 10,500 to 9500 BC, for 1000 years, and melted in 50 years (redrawn from Gray, 1997).

A 10km walk on the old path that leaves the river gorge above Dundonnell, going into the hills along the foot of the An Teallach range, brings you into the beautiful, separate world of Strath na Sealga. There is no road for cars. You can stop on the heights above the deserted strath and look down onto the loch or up into the surrounding glens. Below, on the wide strath floor, close to the rivers running into the loch, are several severely isolated shepherds' cottages alongside their tiny fields, all now abandoned. The river is quite lost in the broad valley, meandering about as if looking for a way to go. Looking south into the corries on the flanks of Beinn Dearg Mor you can see the effects of the melt. During the Younger Dryas, this range was at the farthest northern edge of the great ice-sheet. Glaciers fed from the south filled each valley, pulling rock off the corrie cliff, grinding it up, pushing it down through the ice to the glacier nose. Imagine as the ice suddenly melts. The heap of rock rubble at the front of the ice is quickly stranded, water pours out of the corrie, tumbling with sand and boulders down the hillside into the loch below. It is still there to see. Around the corrie lip are the terminal moraines that once rimmed the edge of the ice. Untidy heaps of heather-covered debris still have the tongue shape of the melting glacier. Rock rubble is strewn everywhere, and down in the loch are the piles of gravel taken up by the meltwater flow or gathered up in dramatic rockfalls, to be suddenly dropped on reaching the loch water below.

The imagination needs little help to recreate this melting chaos as the same Loch na Sealga is approached from the west along the Gruinard River, the loch's present and ice age, westerly directed outlet to the sea. Today, an off-road vehicle track for lazy fishermen follows the south bank of the river up to the lochshore at its outflow. The eight-kilometre walk, on a late autumn day when snow already covers the high peaks, is a walk into dramatic, ice age events. At the start of the track the river is deep and dark, wide and fast in a flat valley between low hills, but changes rapidly to cascading rapids as a tight, cliff-faced gorge is entered. This is the style of the river, repeated several times before the loch is eventually reached. Change of landscape comes slowly on a walk but the slowness itself is dramatic. Passing out of the second gorge along the river the ragged snow-covered peaks of Ben Dearg Mor suddenly appear, a shrouding mist making them seem more menacing and more massive than they really are. It is here that the Younger Dryas ice sheet ended in stiff, glacial fingers. Today, with the mist hanging over the peaks they might still be there, menacingly hidden. They melted long ago of course, but the havoc they created and the debris of the chaos is still there. The path we have followed

Figure 5.9 The misty scenery of Strath na Sealga today. It was the site of catastrophic flooding events during the Younger Dryas.

from the west ends at the lochshore as the view opens to the long, presently tranquil waters of Loch na Sealga. Several wild holly trees grow quietly from the banks. When the ice filled the hills and glaciers filled the corries, there was no tranquillity. As the ice melted, the loch filled and the waters rose. The river became a frightening torrent, the noise overwhelming. The glaciers suddenly detached massive loads of water and rock, causing the loch water to surge. A wave of ice, mud, sand and huge rock boulders plunged through the outflow, squeezing into the gorges, destroying all in its catastrophic path. Today, the mad water is gone, but the chaos of rocks and sand is still there in wide carpets of debris, smooth to the eye today with grazing sheep, but with occasional, incongruous, oversized boulders sticking through. To the scientific observer the original violence is only too evident. Down from the loch, the second gorge is narrow and hemmed in by cliffs. As it opens to a wider valley where the meltwater suddenly lost energy, huge, house-sized boulders have been dumped. To move them needed a monstrous wall of water. Analysing the debris scientifically is difficult. What is left to be seen today is clearly from the greatest of floods, that only imagination can recreate. A walk today along the five miles of the Gruinard River bank to Loch na Sealga creates a nervousness and an uncomfortable realization of what happened during these dramatic events which, remarkably for science, have left unmistakable traces.

In every strath and glen in the north and west of Scotland there are similar marks of the melt. Once recognized they are everywhere: the small burn finding its way through huge boulders; the wider river winding across flat valley floors; mounds of rock and sand in lumpy shapes; tongue-shaped moraines at corrie mouths; sand-strewn beaches. There are many features that have long been observed and described, but they now take on a new significance. The freshness of the features has, of course, always been the indication that they were from the most recent ice event. But if in the past a scientist had suggested that all of these events occurred over a period equivalent to a single human lifetime, it would not have been tolerated. However, this was the case, and with the support of the Greenland ice records, it is time to look at all these things again. There are sceptics, which is only natural. After all, boulder clay and other glacial features were first explained by a kind of biblical catastrophe. This was resisted by dispassionate scientists who showed that the deep, U-shaped valleys, deep corries and the many striking erosional features, as well as the moraines, would have been created only very slowly by ice acting imperceptibly over a long time. This of course is still true; the overall period of glaciation lasted for millions of years. But what influences our present Highland scenery – the rivers, beaches and rocks in the fields – comes from 50 years of melting chaos.

There is no location that shows this better than the village of Achnasheen, the sixth station out of Inverness on the Kyle railway on the way to Skye, and where the train always stops. It is in the very heart of the country at a confluence of three major valleys, part water-, part sediment-filled and drained by the River Bran. Most of the water today flows eastwards to the Cromarty Firth, but the watershed to Loch Carron to the west is not far away and only a few metres higher than Achnasheen station, itself only 150m above sea level, which is why the railway line follows the path that it does to cross the country from one side to the other. It is a unique point topographically; it is also remarkable for its Younger Dryas flood deposits. Around Achnasheen are 20m-high terraces or embankments with remarkably regular slopes and perfectly flat tops that look exactly like the earthworks for a major motorway. They are nothing of the sort, but instead the huge outpourings of boulders, gravel and sand from the surrounding hills, dumped into an ice-dammed lake that covered the valley confluence during part of the final ice melt 11,500 years ago. There is such a volume of valuable building material in these moraines that, inevitably, commercial interests identified the site and wanted to mine it for the construction of cement-type oil rigs in Loch Carron in the 1970s. This did not happen; the terraces are still there and have now been designated a Site of Special Scientific Interest (SSSI). The embankments are 20m high because they were built as deltas into a lake that was 20m deep. The flat floor of the lake is there too,

Figure 5.10
The Kyle of Lochalsh railway uses the bed of an ice age lake to cross the Highlands from east to west. The sketch shows how an ice-dammed lake formed around Achnasheen (modified from Benn, 1996, courtesy of Taylor and Francis).

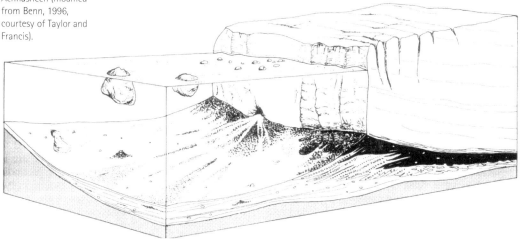

where the station now stands, perfectly preserved because it probably emptied at one go. To the east, where the river now flows, ice and rock dammed the lake close to Achanalt, the fifth railway station out of Inverness, where the train will only stop on request. This barrier was broken suddenly, and massive quantities of sediment were swept eastwards to be dumped in the Cromarty Firth between the Black Isle and Dingwall. The Cromarty Bridge is built on the huge volume of this debris.

Almost as impressive is what happened when the Achnasheen meltwater found the other way to the sea: westwards. It rushed through the narrows at the head of Glen Carron, the way the railway now goes, scouring out a deep gorge. It left scattered moraines along the valley sides and then dumped all the rest of the sand and debris to make the valley floor flat, which stretches all the way from Achnashellac station (the seventh out of Inverness) to the sea, 10km away. The reason why the effects of the flood are so remarkable at Achnasheen is that it was within a deep embayment on the northeastern edge of the Younger Dryas ice-sheet and almost entirely surrounded by local glaciers (Figure 5.8). Once the ice started to melt, the meltwaters from the high ice sheet to the west and south would have been channelled into the embayment, bringing in huge quantities of rock. It is not possible to imagine how catastrophic the 50 years of floods would have been, but the huge volumes of debris and rock rubble that we now see scattered everywhere are a lasting sign of the disaster.

Passing gently through Achnasheen today, in the little two-coach local train on the way to the Kyle of Lochalsh does not stimulate the imagination. Leave the train, leave the station, pass a brass plaque that commemorates all this, and put your face into the biting west wind to walk over the still-fresh debris, like picking your way through dead bodies after a murderous Highland battle.

Similar sorts of terraces can be seen where meltwater torrents carried much debris into the sea on the edges of fjords. These fjords, also known as sea lochs, are where the seawater gradually flooded up glacial cut valleys as sea-level rose at the end of the ice age. The debris was initially dropped in great jumbled piles by melting ice, and then picked up and carried further by the meltwater. These glacial meltwater torrents therefore carried a huge range of sizes of debris. As the water current slowed and then entered the relatively still water of the fjord all the large fragments were dropped into a fan-shaped deposit, with a flat top produced by wave and current action. These are sometimes referred to as raised beaches, as the flat tops are now above the current sea-level. The land rose slowly when the weight of the ice-sheets was removed by melting. Judging by some of the landforms produced in the area, such as Corrieshalloch Gorge at the head of Loch Broom, the water volumes were vast. Ullapool, on the side of Loch Broom, is built on such a raised beach, formed by debris swept in from the Ullapool River and then dumped out into the sea loch.

All these deposits are just the final marks, the graffiti, on a much grander design, built up over millions of years. The Younger Dryas stadial was simply the final episode of the Devensian glaciation, the Devensian in turn being only the last glaciation in the present ice age, which began three million years ago. However, as the Greenland ice records show, ice remained in the Highlands for over 100,000 years and it was this, combined with all the previous glaciations, that sculpted the great erosional features of Scottish scenery. At the last great Devensian maximum, around 25,000 BP, an ice-sheet up to one kilometre thick stretched well into the sea, covering Scotland, most of Wales, some of Ireland and a lot of northern England, although surprisingly leaving a bit of Caithness, the Orkneys, and the tip of Aberdeenshire ice-free. It was this ice that eroded the U-shaped valleys, created the sharp-cliffed corries, ground down the hard rock and left scratches all over ice-smoothed outcrops. There are signs everywhere, even in familiar Assynt. We can return to the summit of Conival, and look again at the boot-breaking traverse of huge blocks of Cambrian Quartzite towards the peak of Ben More itself. We were here first to understand the Moine Thrust. The view is, of course, superb. This time we can look at what the ice has done over the millions of years. The eye is first drawn westwards to the Minches, down the deep glacial groove of Loch Assynt between the solid peaks of Quinag and Canisp. Up against the peaks are the curved cliffs of corries, most facing west, plucked out by the ice as it moved westwards towards the Atlantic during the various ice advances, even though here we are not far from the ice-shed. Twenty-five thousand years ago, at the height of the late Devensian, this view of the wilds of Sutherland would have been like a view in present day Greenland. A huge ice sheet covered the entire area and out to sea across the Minches to the Outer Hebrides. But the twin peaks of Ben More and Conival would still have been visible above the ice. It would have been possible to stand on the peak of Ben More and look down on the rock-strewn surface of the ice sheet a short distance below. There is scientific evidence for this.

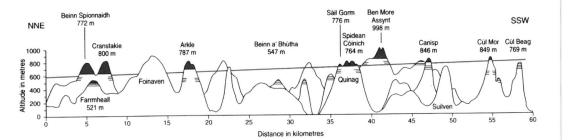

Ben More is 998m (3275ft) high and Conival 987m (3271ft), both Munros (i.e. over 3000ft = 914m). We deduce that the ice sheet here was some 900m (2950ft) thick since all peaks less than this are rounded and smooth from wear under ice. Rock behaves quite differently above the ice to inside it. Within an ice field, in permanent snows, rocks are always below freezing point and it is the movement and pressure of the ice that causes ice moulding and smoothing. Rock above the ice surface is alternately above and below freezing, alternately frozen and melting, depending on the season and the weather. This causes the smallest crack to be exploded by the ice when it falls below 4°C, filled by water when the ice melts and forced open again when it freezes. The rock becomes ice-shattered, breaking into sharp-edged blocks but staying characteristically in place over the mother strata. The Cambrian Quartzite at the summits of Ben More and Conival is just like this; the huge blocks, sharp-edged and heavily shattered, have not moved. Just below these two, on the sister crest of Beinn an Fhuarain, the same outcrops of quartzite were originally below the ice, and they have been either covered with stony moraines or swept clear of debris. Peaks like Ben More and Conival that poke up above the ice sheet are called *nunataks* by the Inuit, their name for bare rock. For them, it is the only rock they see.

The ice leaves many traces of its presence and progress, but those that never cease to impress are scratch marks. At the bottom of 800m or so of ice there are abundant boulders and blocks, trapped in the ice and moving with it. It is these that grind and claw at the bedrock as they pass, leaving lines of scratches on even the hardest surfaces. Arc-shaped cracks are sometimes produced by this process, concave in the direction of ice travel.

In the Assynt area, the directions of ice flow can be beautifully mapped by measuring the abundant scratches on the Cambrian Quartzite surfaces. Usually, delicate ice scratches are weathered away quite rapidly from most surfaces, but the quartzite is so hard that

Figure 5.11
Not all the Highland mountain peaks were covered by the last great ice sheet; some poked through the surface. From Ben More Assynt, it would have been possible to look down on the ice. Dark blue = nunataks, Red line = ice surface. (Modified from McCarroll et al., 1995, by courtesy of Taylor and Francis.)

Figure 5.12
Below one kilometre of ice, even the hardest rocks show ice scratches indicating in intimate detail how the ice moved 25,000 years ago (after James Geikie, 1894).

it resists weathering, although of course it was not hard enough to resist being scratched below the ice. All the marks are from the Devensian and probably between 35,000 and 25,000 years old. The Younger Dryas ice did not reach these places, as can be easily deduced from the moraine stranded higher up the valley floors. Follow the scratches on the bare, seemingly polished quartzite on the dip slope of Canisp, the flanks of Conival, and wherever they can be seen. The movement is, of course, west, but hills clearly got in the way, or there was a local source of ice flow, so directions varied. Map out all the scratches and intimate details become clear. The ice sheet moved along Gleann Dubh, over the top of Stronchrubie cliff, around Canisp, through Loch Assynt and along the flanks of Suilven. Occasionally, as on the slope of Canisp, there are lines of scratches that defy logic, one criss-crossing another, but logic is probably not the right tool to use under 800m of ice. However, such an intimate monitoring of the ancient ice shows the strength of its influence.

The scratches are easy to understand, but was there something else going on under the ice? Large-scale structures on the surface of the Earth are not always easy to see, and until relatively recently we could only look from close up. Small-scale structures can be seen by eye in their entirety, but otherwise, techniques must be used to picture what they look like. For example, folds that are miles across can be mapped out by taking many measurements of the amount of slope of the rock layers as they appear at the surface. The use of aerial photographs and now, better still, satellite images, allow us to stand back and enhance the view from a distance to see large-scale patterns (Figure 5.13).

There are not just small-scale scratches on the surface of some glaciated rock. The quartzite, and schist in particular, have patterns of deep grooves scoured across them. The photograph below shows the land's surface as viewed from a satellite using reflected laser beams that allows us to see the shape of the rock surface through any peat or recent deposits. At first sight the surface looks covered in scratches until you look at the scale of them. These are indeed large, typically from many hundred metres to kilometres in length, and tens of metres across. The term megagroove has been used for them. They appear to be cut beneath a fast moving part of the ice sheet in a process not yet fully understood. Perhaps debris being carried by high-pressure water has a part to play here, rather than just mechanical scouring as with the striations. Research is currently being carried out to clarify the process that took place.

One force that the ice exerted that is not immediately evident, and was only eventually discovered by the older workers, is that of its sheer weight. The scratches and erosion are clear and the moraines obvious, but the weighing down of the land was less easy to realize. We are familiar now with the fact that the huge weight of an ice-sheet 800m thick will depress the land. In human spans of time rock is solid, but over geological time it will behave like a very sticky, viscous fluid (as described in the next chapter). Put a weight on the land and it will sink; take it away and it will move up again; so-called isostatic adjustment. At present, the parts of Scotland which were beneath the thickest ice 25,000 years ago are still rebounding, or lifting upwards, at perhaps 0.25cm each year, or 15cm in a lifetime. Not much perhaps, but like all things geological, over thousands or millions of years, very significant. This is best seen at the edges of the land where the sea acts as a measure for its movement. Around Gruinard Bay, where there is a beautiful, clean sandy shore, there are terraces six metres above the present beach, as at Ullapool. This is where the sea-level stood during the Younger Dryas flood. Since then, however, the land has rebounded upwards more than the six metres of the terraces, because at the same time as it has been rising, the sea has also been rising, ice continuing to melt in the Arctic and Antarctic. The land has obviously risen faster than the sea to leave the terraced beach stranded. It is not just at Gruinard Bay that there are sea terraces; they occur all around the coasts and from them we can map out the rising of the land, which in turn confirms where the ice would have been, and how thick. The highest rises are just north of Glasgow, Gruinard Bay being in the zone of moderate rebound. Further to the north where there was no ice, there is no rebound and the land is being slowly drowned. Only

Figure 5.13
Satellite image of the land surface just north of Ullapool, showing grooves cut many kilometres in length. Note the scale (Reproduced with permission of ELSEVIER BV).

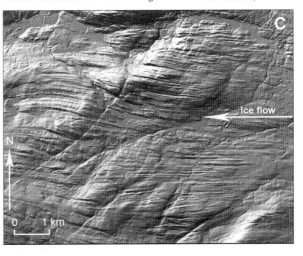

10,000 years ago Orkney consisted of one large island, instead of the myriad of islands and skerries today, which is a sign of rising sea-level.

From the raised beaches, the moraines, the U-shaped valleys, the tumbling burns and the glacial scratches, northern Scotland cannot hide its links with the ice. Because of this, it is often forgotten that heat, as well as cold, shared in Scotland's past. The present climate is not one of northern Scotland's best selling-points. The evidence of past warmth is much less obvious than of cold, but evidence there is, especially in the distribution of plants, the growth of trees and the history of insects and animals. So where are we today? Waiting for an ice age or waiting for the climate to warm up? 'The present is the key to the past', according to geologist Charles Lyell. The future as well?

The present reality is that we are now 11,500 years into a warm, interglacial period. That is how long it has been since the Younger Dryas melt. The Greenland ice cores indicate that temperature was highest immediately after the end of the Younger Dryas, and has been gradually cooling since 9500 BC as we head into another ice age. Vegetation records from lake sediments in the Highlands indicate the same: a warm peak soon after the Younger Dryas, followed by a long-term temperature decrease. Pollen from trees blows into the Highland lakes each year and is preserved in the layers of mud and silt on the lake bottom. The dust-sized pollen of hayfever fame can be easily extracted from the mud layers and, when enlarged 400–600 times under the microscope, can be analysed to show how tree species changed and the abundance of grasses varied. Changing species diversity is a measure of changing temperature and climate in general. Chris Smout writes:

> There was a time after the final retreat of the last ice sheet
> from Scotland when invading trees covered all the land
> that was neither bog, mountain top nor fresh water loch,
> with primeval forest.

There are still very rare stands of this famous 'Great Wood of Caledon' that covered Scotland in the period of warmer climate: now it is only the grasses that survive. The Great Wood is very distinct in the pollen records as a mix of pine, birch, hazel and oak, with grass pollen very much diminished. The forest thrived from around 6000 to 3000 BC with its greatest extent at the earlier time. It then diminished slowly and died out as the climate deteriorated, becoming gradually smothered by the blanket bog familiar to all who walk the hills today. Those acid-bleached pine stumps (*Pinus sylvestris*), that often appear out of the peats in the gullies eroded by the boots of walkers on high mountain slopes, even to over 750m up, are from the long-dead forest and quite likely to be 5000 years old or more. Remarkably, in Fortingall village in Perthshire, one tree still lives from this time: the famous Fortingall Yew. Although not really part of the forest, it is 5000 years old and, from a very recent, sophisticated electrical induction survey, parts of it are known to be flourishing. There is a puzzle here, however: glance at a map of the Highlands and it is covered with

Figure 5.14
There is so little left of the 'Great Wood of Caledon' that once covered the Highlands. Its absence today is a result of climate change and man's influence.

forests. In the west for example, Dundonnell Forest and Glencanisp Forest, Abernethy and Rothiemurchus Forests on Speyside, or the Forest of Atholl north of Perth. They run into the hundreds and may be called forests, but not from the presence of trees: there generally are none. 'Trees! Wha' ever heard tell o' trees in a forest?' is a nice Scottish comment. In the Highlands it is a legal but not an ecological term. They are actually 'Deer Forests' and from 1860 onwards had hunting rights with restrictions as to grazing, woodcutting, building and so on. The term has a similar sort of status to that of 'green belt' in England. The confusion in the Highlands is, of course, that these had indeed been forests in the past – the magnificent and much missed Great Wood of Caledon, now only a vestige in the peat. The tiny stands of real forest at Abernethy and Rothiemurchus are only a forlorn reminder. Sadly, the last few thousand years of Highland history are in its bogs, a history of increased wetness and persistently lowering temperatures. These changes since the ice melted, confusing as they may seem to us, are not exceptional; rather they are to be expected. Records from deep-ocean cores show that regular warm-to-cold cycles have been repeated during the last three million years. Each cycle, especially the most recent, starts with the highest temperatures and ends with the coldest and ice, and lasts between 100,000 to 150,000 years. Warm always leads eventually to ice, and then suddenly warm comes back again and the ice goes. We have only described the results of the latest, and it is just the latest, sudden warming that ended the Younger Dryas. There were many ice events before, and many periods of melting. It is even the case that the temperature cycles are predictable, caused by the way that the Earth's orbit and spin change regularly in relation to the Sun under the gravitational pull of the other planets, leading to variations in the intensity and distribution of sunlight reaching the Earth's surface. There are three types of regular change, known as Milankovitch cycles, after the Serbian scientist who calculated them in the 1920s and 1930s. The periods of the three cycles are roughly 20,000 (precession), 40,000 (tilt) and 100,000 (eccentricity) years. Precession is the way the Earth's axis of rotation wobbles like a spinning top. Tilt is the change in the angle of the Earth's spin axis in relation to its orbit; it is at present 23.5°. Eccentricity is the measure of the Earth's orbit about the Sun, which varies from a circle to an ellipse. Each variable is independent of the others, but together they combine to create a cyclic signal of variable solar energy, which accounts for nearly 80% of long-term climatic variation. If this is true of the past, one assumes it will be true in the future as well. Indeed, using Milankovitch cyclicity to predict the coming climate, one researcher states bluntly:

We see cooling for the next 5000–6000 years, another cool maximum at 24,000–25,000 AP (after present) and then an apparent glacial maximum at around 59,000–60,000 AP, followed by deglaciation.

This is the fashion of scientific prediction, chillingly delivered in a toneless telephone time machine voice. 'There will be a new ice age. It will begin 25,000 years from now'. Calculations do vary, and some workers give us a little longer than this, as the illustration shows (Figure 5.15), but all agree a new ice age is coming.

Predicting a new ice age is like predicting the seasons: they come and there is no surprise. The surprise, of course, is that they are always different: a colder winter, a wet spring, a hotter summer or no summer at all. We do not ask the meteorologists to predict that winter will come; we want to know what kind of winter. We do not ask whether day and night will come; we want to know what kind of a day, wet or fine. These are the hard predictions. An ice age is certainly coming. Milankovitch gives us another 25,000 years or so before it begins, but in the meantime, the decades and the centuries are not predictable and they may be hotter or may be colder. There is even room for global warming! In order to predict shorter-term changes that will affect our

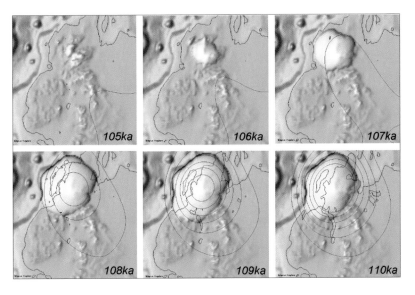

Figure 5.15
From the behaviour of climate and ice activity over the last million years, it is possible to predict the arrival and development of the next ice age. Ka = 1000 years AP (after present) (after Magnus Hagdorn, 2003. PhD thesis, Edinburgh).

immediate generations, climatologists are desperately searching for short-term cyclicity, repetitions over hundreds or even tens of years, rather than the ten thousands of the Milankovitch repetitions. In this way it will be possible to use the past to predict the future, simply extending the regular repetitions beyond the year zero. So let's have a look at the shorter variations over the last few thousand years, especially in the Highlands. These are going to be found by studying recent natural events, human history and actual instrument records. Written accounts are now in range. Let's look at the last 3000 years, 1000 BC to the present.

The endless heather of the bare Highland hills is recent, but when and how all those magnificent forests disappeared is disputed. As we approach today, the effect of human activity becomes intertangled with natural effects, and the environmentalist's adage is 'More people lead to less forest'. In fact the forests began to go under the bog in the Neolithic, around 3000 BC, before Man could have had an effect. The Neolithic (the new Stone Age), starts in Scotland somewhere around

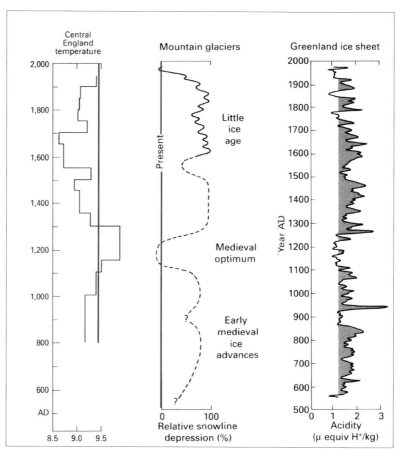

Figure 5.16
The climate over the last 1500 years has been variable, and continues to be so today. A comparison between measured and inferred temperatures, mountain glaciers and the record of the Greenland ice sheet (modified from Porter 1986).

3500 BC, continuing until about 2000 BC, followed by the Bronze and Iron Ages, the latter overlapping the Roman presence in the south and continuing to 1000 AD. The population of Scotland at around 7500 BC (Mesolithic) is guessed to have been as little as 60, but the evidence for an accurate guess is very sparse. Agriculture and associated tree clearance did not arrive until 2000 BC, although established population centres did exist before that. There is one on the Ord by Lairg at the southern end of Loch Shin, dated at 2500 BC. There is even a settlement of the same age in one author's back garden. Despite this, even by 1000 AD, there were perhaps only 250,000 people in the whole of Scotland, with very few in the far north and mainly limited to the coasts and lochshores, settlers being unable to penetrate the Great Wood of Caledon. But they did penetrate the forests eventually, and by Man's hand or by climate, the trees disappeared. The end was rapid. By the 1400s woodland was beginning to be legally protected. However, the human population, especially of the Highlands, was never higher than in the century 1750–1850 AD, and during the Highland Clearances of the early 1800s cutting wood was forbidden, as trees had become so scarce. So it was not the sheep that cleared the glen – they were only introduced in large numbers just before 1800 – though they certainly kept it so (and, with deer, still do today). To quote yet another of the understandably emotional surveys of the effect of sheep in the Highlands:

> *The transformation of Moidart and Arisaig into a short-term productive source of cheap wool and meat for urbanizing Britain (in the 1780s and 90s) was rapid and devastating.*

Left alone today, trees would colonize many parts of the Highlands. Small exclosures constructed by conservation organizations such as Scottish Natural Heritage show this to be true. The small, isolated islands in many lochs show what the natural vegetation would be like: Scots pine, birch, hazel and sometimes oak. The National Nature Reserve at Creag Meagaidh, above the north shore of Loch Laggan, has even allowed the regrowth of this natural vegetation over a large area to be observed over the last 20 years.

In the past, when the trees disappeared, so too did the native animals. The large, romantic ice age mammals such as the mammoth, sabre-toothed tiger and woolly rhinoceros had mysteriously died out during the Devensian glaciation. However, initially at least, the Great Wood of Caledon contained a varied range of mammals including red deer, roe deer, great elk, wild horse, wild boar, beaver, wolf and brown bear. The Romans sent a Caledonian bear to the circus in Rome, although these abused animals managed to last until the tenth century. The beaver lasted until the mid-sixteenth century (which, incidentally, tells us that deciduous forest near water was still common until then). Now the beaver is being reintroduced in selected areas to diversify the habitat. The last wolf in Scotland was famous and various. It was killed near Killiecrankie in 1680 by Sir Ewan Cameron, but also in Sutherland around 1700 by hunter Polson, then eventually by Eagan Macqueen in 1743 near Inverness (but only perhaps). There are still local sightings of large, skulking, shadowy animals, especially after pub closing time on winter evenings. At least man's culpability in the extinction of the wolf is clear even if the exact date is not. Man's meddling with the environment makes the detection of climate change before early medieval times quite difficult.

Historical records make things easier, and from 1700 AD continuous climatic records are available from Edinburgh and elsewhere.

Those intrepid Victorian geologists Ben Peach and John Horne had a part in finding animal remains from inter-glacial periods in Durness limestone caves in Assynt, whilst they were working on the other rock conundrums. Quite appropriately they have been named the Bone Caves. Assynt has the largest cave systems in Scotland and more are being discovered all the time. Animals living on the surface in inter-glacial periods usually have their remains scavenged on, including the bones. This is quite normal in most land environments, where anything worth eating doesn't go to waste. If, however, the animal dies in a cave, or its remains are washed into a cave then its bones are likely to be buried and preserved, as was the case in the Bone Caves. Bones have been found from wolves, brown bear and artic fox, with skulls identified as being from a polar bear and a lynx. Human remains have also been found but little is known about them, or indeed about a 2000-year-old walrus ivory pin from the Iron Age.

However, even when actual temperature and rainfall values are known, what can be appreciated better are the associated human activities. For example, high temperatures in the past are illustrated by the fact that the Romans grew grapes in the Thames valley around 500 AD, and that the Domesday Book records 38 vineyards in the south of England in 1086 AD. The latter warm period, from about 1100–1350 AD, is well enough known to be called the 'Medieval Optimum'. Since vineyards are growing again in the south, we are perhaps now in the

'Modern Optimum'. If there were warm periods in the past, there were also cold ones. The 'Little Ice Age', between 1680 and 1850 AD, was markedly cold. The lovely old oil pictures of the Thames frozen over in London, and skaters around braziers roasting chestnuts in the middle of the river are a witness to it. Those of us who live in the Highlands often wonder why curling ponds are kept in the villages; they never freeze today. In Norway, where there is still permanent ice a short way into the mountains just north of Bergen, there is a different but poignant record. A vivid report made in 1744 by the vicar of a parish on the edge of the ice gives real meaning to the phrase 'ice age'. It reads:

> *Its colour is sky blue and it is as hard as the hardest stone ever could be with big crevasses and deep hollows and gaps all over and right down to the bottom. Nobody can tell the depth although they have tried to measure it. When at times it pushes forward a great sound is heard, like that of an organ and it pushes in front of it unmeasurable masses of soil, grit and rocks bigger than any house could be, which it then crushes small like sand. In summer there is an awful cold wind blowing off it. The snow which falls on it in winter vanishes in summer but the ice glacier grows bigger and bigger.*

In Scotland there were no glaciers, but naturalist and traveller Thomas Pennant wrote in 1771 that 'Snow lies in the chasms of Ben Wyvis in the form of a glaciere throughout the year': not the case today but perhaps the origin of the present tradition that a snowball can be collected off the hill in any month of the year. More intriguing, though, is the suggestion that there were no Highland midges (*Culicoides impunctatus*) between 1750 and 1850. General Wade and his successors built their military roads without getting bitten. Boswell and Dr Johnson, who famously toured the Highlands in 1773, were apparently not hindered either, while hunters in the early 1800s speak of idyllic days on the moor, with abundant wildlife but no midges. One change that has not yet been played on the environmentalists' trumpet is the decline in heather – up to 40% of coverage in some places – bracken is the new menace. The most recent threat to upland bog, however, is from wind farms, because they require huge road networks in this most fragile (and supposedly protected) environment. Climate can be blamed for some things, but some of men's unnecessary activities are more than an equal agent.

Figure 5.17
The large romantic
animals shown in
cave paintings died
out mysteriously
during the Devensian.
The Irish Elk (after
Goldsmith, 1848).

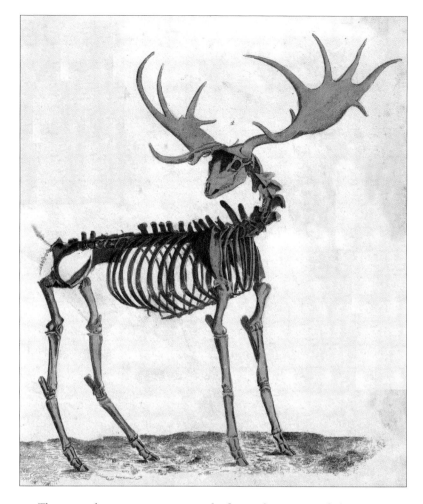

Figure 5.17
The large romantic animals shown in cave paintings died out mysteriously during the Devensian. The Irish Elk (after Goldsmith, 1848).

The actual temperature records from the years of these stories show that they are true. Measured and estimated temperatures, correlated to the Greenland ice cores, show a distinct warming from about 1050 AD to its warmest between 1150 AD and 1300 AD, the 'Medieval Optimum', while temperature drops to a marked low between 1650 AD and 1850 AD, the 'Little Ice Age' (Figure 5.16). Instrumental records, which begin around the 1700s, show that there are changes in the climate over tens and hundreds of years. For example, records of changes in the growing season, linked to measured air temperatures in central England since 1680, show constant variations of considerable amplitude. In fact all instrumental records show that variation is normal; we cannot expect consistency even if we wish it. Although the changes recorded by early instruments since the 1680s, by detailed modern measurements since the 1960s, and by satellite measurements since the 1970s have been much analysed, no definite cyclic pattern has yet been found in them; there is no mini-Milankovitch. Recognizing cyclicity, as we have seen, allows future trends in the climate to be predicted. However, it is as well to realize that the difference between the historical 'Medieval Optimum' and 'Little Ice Age' is only 1.0°C at most, whereas the changes being charted from the past during the actual ice ages in the Highlands are between 5.0 and 8.0°C. In fact, the present interglacial has been more stable than is normal, and for longer. We are now enjoying the best climate in 120,000 years. Be this by man's doing, or natural, we don't yet know, but it will not last. There is a coming ice age.

More menacing, though, are the massive and abrupt climate changes recorded during the ice ages and glaringly unexplained. The reasons for them are somewhere in the climatic system, and we have not yet found them. To science today, they are quite unpredictable. Being obsessed with global warming is like giving Nero sheet music: Rome still burns.

Further Reading

Books, Pamphlets

Bell, M. and Walker, M.J.C. 1992. *Late Quaternary Environmental Change.* England: Addison Wesley Longman Ltd. pp.273. ISBN 0-582-04514-2

Bennett, M.R. and Glasser, N.F. 1997. *Glacial Geology.* New York: John Wiley & Sons. pp.364. ISBN 0-471-963453-3

Gordon, J. (Ed.) 1997. *Reflections on the Ice Age in Scotland.* Scottish Natural Heritage & Scottish Association of Geography Teachers, Glasgow. pp.188. ISBN 0 9524210 11

Gourlay, R. 1996. *Sutherland, an Archaeological Guide.* Edinburgh: Birlinn Ltd. pp.105. ISBN 1 874744 44 0

Grove, Jean M. 1988. *The Little Ice Age.* London & New York: Routledge. pp.498.

Houghton, J. 1997. *Global Warming, the complete briefing.* Cambridge, UK: Cambridge University Press. 2nd Edition, pp.251. ISBN 0 521 62089 2

Price, R.J. 1983. *Scotland's Environment during the last 30,000 years.* Edinburgh: Scottish Academic Press. pp. 224.

Ramsay, P. 1998. *Revival of the Land, Creag Meagaidh National Nature reserve.* Scottish Natural Heritage. pp.111. ISBN 1 85397 238 X

Smout, T.C. (Ed.) 1993. *Scotland since Prehistory. Natural change and human impact.* Scottish Natural Heritage. pp.140.

Taber, C.A.M. 1896. *The Coming Ice Age.* Boston: Geo. H. Ellis. pp.94.

Trewin, N.H. (Ed.) 2002. *Geology of Scotland.* Geological Society London. 4th Edition, pp.576. ISBN 1 86239 126 2

Scientific Papers

Alley, R.B. and Bender, M.L. 1998. Greenland Ice Cores: frozen in Time. *Scientific American,* Feb. pp.66–71.

Alley, R.B., Meese, D.A., Shuman, C.A., Gow, A.J., Taylor, K.C., Grootes, P.M., White, J.W.C., Ram, M., Waddington, E.D., Mayewski, P.A. and Zielinski, G.A. 1993. Abrupt increase in Greenland snow accumulation at the end of the Younger Dryas event. *Nature.* Vol. 362, April. pp.527–529.

Benn, D.I. 1996. Subglacial and subaqueous processes near a glacier grounding line: sedimentological evidence from a former ice-dammed lake, Achnasheen, Scotland. *Boreas,* 25. pp.23–36.

Bennett, M.R. & Boulton, G.S. 1993. A reinterpretation of Scottish 'hummocky moraine' and its significance for the deglaciation of the

Scottish Highlands during the Younger Dryas or Loch Lomond Stadial. *Geological Magazine* Vol 30. pp.301–318.

Dansgaard, W., Johnsen, S.J., Clausen, H.B., Dahl-Jensen, D., Gundestrup, N.S., Hammer, C.U., Hvidberg, C.S., Steffensen, J.P., Sveinbjörnsdottir, A.E., Jouzel, J. and Bond, G. 1993. Evidence for general instability of past climate from a 250-kyr ice-core record. *Nature*. Vol. 364, July. pp.218–220

Krabbendam, M., Bradwell, T. 2014. Quaternary Evolution of glaciated gneiss terrains; pre-glacial weathering vs. glacial erosion. *Quaternary Science Reviews* 95, pp.20–42.

Lawson, T.J. 1996. Glacial striae and former ice movement: the evidence from Assynt, Sutherland. *Scottish Journal of Geology*. Vol. 32 (1). p.59–65.

McCarroll, D., Ballantyne, C.K., Nesje, A. and Dahl, S-O. 1995. Nunataks of the last ice sheet in Northwest Scotland. *Boreas*. Vol. 24. pp.305–323.

Petit, R., Jouzel, J., Raynaud, D., Barkov, N.I., Barnola, J.-M., Basile, I., Bender, M., Chappellaz, J., Davis, M., Delaygue, Delmotte, M., Kotlyakov, V.M., Legrand, M., Lipenkov, V.Y., Lorius, C., Pépin, I. Ritz, C., Saltzman and Stievenard, M. 1999. Climate and atmospheric history of the past 420,000 years from the Vostok ice core, Antarctica. *Nature*, Vol. 399, June. pp.429–436.

Porter, S.C. 1986. Pattern of forcing of Northern Hemisphere glacier variations during the last millennium. *Quaternary Research*, 26, pp.27–48.

Shennan, I., Rutherford, M.M., Innes, J.B. and Walker, K. 1996. Late glacial sea level and ocean margin environmental changes interpreted from biostratigraphic and lithostratigraphic studies of isolation basins in northwest Scotland. In: Andrews, J.T., Austin, W.E.N., Bergsten, H. and Jennings, A.E. (Eds) *Late Quaternary Palaeooceanography of the North Atlantic Margins. Geological Society of London*. Special Publication 111. pp.229–244.

Stoker, M., Bradwell, T., Howe, J., Wilkinson, I., McIntyre, K. 2009. Late glacial ice-cap dynamics in NW Scotland: Evidence from the fjords of the Summer Isles region. *Quaternary Science Reviews*, 30 pp.1–24.

Willemse, N.W. and Tounqvist, T.E. 1999. Holocene Century-scale Temperature Variability from West Greenland Lake Records. *Geology*. Vol. 27, No. 7. pp.580–584.

Chapter 6   LEFT-OVERS

The Lewisian Gneiss and the creation of continents

A desolate place to stop the land, deserted buildings with boarded up doors and blocked out windows, broken dykes, bits of half-buried, unrecognizable, rusted machines scattered around like a lost battle. A white-painted, computerized lighthouse on the edge of high, dark cliffs permanently besieged by a heaving, grey-streaked, wind-bullied sea. The buildings are a sign of human activity but there are no humans, except for a straggling group of tourists huddled with their hoods up in the lee of one of the deserted blocks, trying to keep out of the tearing wind and waiting pathetically to be taken away back to comfort. This place doesn't want people: it is plain uncomfortable. This place is Cape Wrath on a good day. It is neither the most northerly part of the Scottish mainland nor the most westerly; it is where the land ends and is, simply, the most desolate.

The image of Cape Wrath has perhaps mellowed a little in recent years with the opening of the Ozone Café. There is no doubt that a human face and the offer of hospitality softens even the harshest of environments, and it was opened by none other than the Princess Royal. She too appears to enjoy these rugged and testing places. Unfortunately that welcoming face was not always there for the owners themselves. John Ure was left stranded once whilst his wife went for the Christmas shopping, and she didn't make it back with the turkey until after Christmas! Such are the extremes of weather in this part of the world. The lighthouse is the end (or indeed the start) of probably the toughest long-distance footpath that this country has to offer. The Cape Wrath Trail runs from here to Fort William through some of the wildest country, and avoiding most of the habitation that exists.

The narrow waters of the Kyle of Durness estuary at Keoldale separate the 18km track to Cape Wrath from the mainland; you cannot drive there in your own car. On his first visit one author cycled the 35 bumpy kilometres. On a good day the small bucket of an aluminium ferry boat can take a dozen passengers and a bike or two from the mainland to the couple of uninhabited houses on the far side of the Kyle. It takes about 10 minutes. A minibus lift (or a bike) then takes the same passengers the 18 kilometres to the Cape. But if there is a bit of wind or the tide is too low, too fast or too – tidal,

Figure 6.1
Garvie Island just
west of Cape Wrath,
feeling the attention
of aircraft fire power
(courtesy of Gill Robb).

or there was too much drunk the night before, the ferry does not go. There are also other reasons for it not going, like on the day of the local Highland Games, a local birth or a death and funeral. Or even too much drunk the night before – again. But there are days when it does go. The ferry across the Kyle is part of the official route to the only sea-to-land military firing range in Europe, owned by the MOD. NATO forces combine land, air and sea manoeuvres and fire various ordnance. Garvie Island (Figure 6.1) receives special attention from aircraft. You may think the photograph shows an eruption, or indeed, a meteorite impact, but as with much of the destruction on the surface of our planet, it is inflicted by ourselves. The dour desolation of Cape Wrath seeps into this hinterland and there are no trees, no wildlife or colour, only endless, dismal, bomb-cratered bog – and the occasional cyclist.

Desolate and disagreeable as it may be, Cape Wrath has a special place in the history of rocks, and a visit is worth any discomfort for a geologist. Walk a little south of the Cape and, if the wind permits, continue down towards the rock edge to be in view of the high cliffs just below the lighthouse. The large geological structures across the full height of the cliff face are quite clear and the layers can be seen to have been squeezed into heavy folds like a huge theatre curtain dropped on the floor. Small folds ruckle the insides of larger ones. The grey and white banding picked out on the weather-etched surfaces at the cliff top show even greater detail; the bending and squeezing go right down to the finest layer. It is not these structures that give the rocks their geological importance; they are just a symptom of it. What is important is their great age, which is why they feature in this chapter. They have had time to be squeezed and folded because they are three billion years old and are Precambrian, Lewisian Gneisses. These rocks are old enough to have existed through two-thirds of the Earth's entire existence: an incredible length of time. In this crumpled cliff is a rare piece of the extremely distant past. Observed, studied and analysed in the right way, these rocks are full of time. This is the subject of this chapter – squeezing that time out of these gneisses and extracting the story of their origins and their existence.

The 'Lewisian' as it is colloquially called, and what Dr John Macculloch first called it in 1819, is geologically a gneiss (Chapter 1). The name Lewisian comes from the fact that it makes up most of the rocks of the Outer Isles (Lewis being the main island), although it is equally found on the mainland; and gneiss, a general geological term for a banded, coarsely crystalline rock considered to have formed at very high pressures and temperatures, being the product of 'high-grade metamorphism' and created deep below the Earth's surface. Peach and Horne described the Lewisian Gneiss as 'fundamental' and everything in its nature, structure, stratigraphy and detail clearly shows it to be the oldest rock in the area. But the early workers were not to know that it is fundamental indeed, even in terms of our entire planet's development, and not just fundamental in terms of Scottish

Figure 6.2
Lewisian Gneiss wrinkled by tight isoclinal folds inherited from its time nearly 30km below the surface at very high temperatures up to 3.0Ga ago. The same isoclinal structures are found at scales of 10s to 100s of metres. Pencil 15cm.

geology. They would have been astonished to discover the three-billion-year-old dating, an age far older than anyone imagined at that time (1907), the expected figure being only of the order of hundreds of millions. Lord Kelvin had actually calculated the age of the Earth at 98 million in 1861, from the rate at which it was (and still is) losing heat. A nice idea, but he got his sums wrong. Two hundred years before this the Earth's age had been estimated at even less, a mere biblical 6000 years (Bishop Ussher's famous creation at 9.00am, Monday October 23rd, 4004 BC, in other words – first thing one Monday morning), so Kelvin was actually making good progress. Today, astronomical and geological ages tally and the Earth, the major planets, and especially the meteorites, are all as old as each other. Carbonaceous chondrite meteorites, which represent what is left of the primitive matter that coagulated to form the Solar System's planets, are dated at 4.56Ga. Almost all the dated rocks brought back by the Apollo missions have ages between 3.2Ga and 3.8Ga; one sample, however, gave the dating of 4.45Ga, the age of the Moon. The Earth's astronomical age is 4.56Ga, the same age as the meteorites. So why is the Lewisian Gneiss of Cape Wrath, 'only' 3.0 billion years old, so special compared to the much older rocks from the Moon or the meteorites?

For Mercury, the Moon and Mars, a cratered, pock-marked surface is their face, familiar to us all now from those remarkable space photographs. Pock-marked, that is, on all these planets but not on Earth. Indeed, so pock-marked is our Moon that pock-marks are pock-marked, small craters on big ones. A map of the Moon is a map of its craters, each one the record of a bombardment from space, of a rain of meteorites (the craters are not volcanoes, as was thought until the 1940s, as described in Chapter 4). In fact the Moon has been so pulverized that its outer layer is often just dust, the regolith, a word created to describe Moon 'soil'. The age of a pock-marked planet's surface can be calculated from its craters; the more there are (and bigger), the older it is, the Moon's surface being very old, 3.8–3.2Ga (even 4.45Ga). And yet here on Earth this does not work; there is just one famous crater, *the* Barringer Meteorite Crater in Arizona (which Mr Barringer owns, so you cannot visit it) and a few others that are only now beginning to be known (which you can visit). They are a rare and insignificant surface feature, although experts say that there are in fact about 200 craters world-wide, of all sizes and all ages, but usually difficult to see. There is no crater to be seen for some impacts, such as the one off the coast of Mexico that marks the end of the reign of dinosaurs, or the one that struck the north of Scotland 1200Ma ago and produced the ejecta mentioned in Chapter 1.

A map of the Earth is a map of continents and seas, mountains and coasts; over millennia the map changes, oceans change shape, continents move, mountains rise and are eventually weathered and eroded away. It is not a map of craters. A great deal of the surface of the Earth is geologically young, forever renewing itself like human skin, although adding a few new wrinkles here and there as time goes by. No rocks below the oceans are older than 200Ma and most are younger than 60Ma, the North Atlantic Oceanic rocks described in Chapter 4 being typical. Old rocks only occur on the continents, and as the Cape Wrath Lewisian outcrops show, parts of these can be very old indeed, 3.0Ga or more, almost (but not) as old as the Earth itself. How can this be – very young ocean, very old continent? The Lewisian can tell us.

But first, just to set the scene, let's have a look at the Earth's geological 'headlines', the major events on our planet that we know about, and then look at what these imply, before learning more about the Lewisian and perhaps thinking about returning to Cape Wrath. Hmmm!

The age of the Earth is 4.56Ga. The oldest known rock on the Earth's surface is 4.03Ga, the Acasta Gneiss in northwest Canada. The prize for the oldest sediments, and perhaps the oldest remains of life itself, is claimed by the Isua Gneiss in Western Greenland, which is just marginally younger at 3.9Ga. The oldest known solid Earth objects are minerals, tiny zircon crystals a few millimetres long from Jack Hills in Western Australia, with ages up to 4.4Ga and much older than the sedimentary rocks containing them. Which is rather like decoding the DNA of a blood stain and fantasizing about a murder; it could equally have been a nose-bleed. In addition, these same tiny zircons, when examined for their oxygen isotopes, are said to indicate

Figure 6.3
The major events in the Earth's geological history over the vast time of 4.56Ga. Complex life only began at 0.54Ga. PH = Phanerozoic.

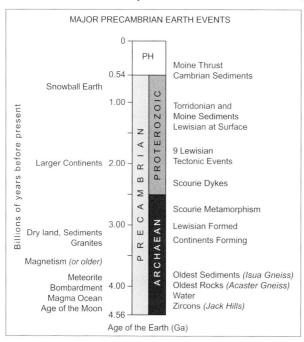

that even at this very early date only 150Ma or so after Earth's beginning, water existed.

Perhaps. The oldest known widespread sediments (also in western Australia), which certainly indicate weathering by water on 'dry' continents, are much younger at an age of 3.2Ga. The oldest granites, indicators of 'true' continental crust, are equally 3.2Ga but only common much later. The Lewisian Gneiss, remember, is 3.0Ga old, so truly part of cradle Earth. It is always a risk when using the shorthand 'Ga' that the great lengths of time being invoked will not be realized. It took less than 0.4Ga for humans to evolve from fish (400 million years, Chapter 3), so 1 billion years, 1Ga, really is a vast period of time. The bits of the Earth we are most familiar with, that are described in Chapters 1 and 3, are those containing body fossils, and are younger than 'only' 540Ma (Cambrian), that is, only from the last 12% or so of Earth history. Since the Lewisian was formed, on the other hand, the Earth has passed through more than 60% of its development. Through two-thirds of the Earth's billions of years of planetary history the rocks of Cape Wrath have survived. They have been folded, squeezed, heated, fractured, re-crystallized and heaved upwards certainly, but they have survived. Being more than 2.5Ga old, they are geologically called Archaean, and there are not so many like them to be seen as outcrops, because as rocks get older on Earth they get rarer. Exposed early Archaean rocks, that is, those older than 3.5Ga, are very rare indeed. This somehow sounds logical, perhaps because it is so with people, but with planets it is the opposite. Planetary surfaces are generally very old, which is why they are so pock-marked. So what on Earth is happening?

It all comes down to heat, as is so often the case in astronomy, in the past as now. Dr James Hutton, who features in the next chapter, thought in 1795 that the Earth's heat came from a continuous layer of coal burning below the surface: fun to imagine but wrong, of course. However, he was quite right to evoke heat at the origin of the Earth engine. The Earth and all the other planets came from a swirling disc of dust that condensed into larger and larger lumps, each bumping into the other and fusing and eventually forming the planets. As bigger and bigger bits came together they acquired heat from the friction of formation and eventually, when planet-sized, from gravity and continuing bombardment. Hot planets were formed from cold space dust. The Earth also heats itself from internal radioactive decay, a source more active in the past but still at work now. Despite this, the Earth today loses twice as much heat as it can create, that extra amount coming from its hot start. Slowly and persistently the Earth is cooling down in absolute terms and, as Lord Kelvin rightly calculated, without the radioactive heat source, which he did not know about, it would long ago have cooled down to be as stiff and cratered as the Moon and Mars. But as we all know, deep mines are hot. The deeper the mine the hotter, and in the South African gold mines, at depths of five kilometres, this means temperatures of up to 70°C, only bearable for the miners through being cooled by refrigerated air. The Earth's interior, therefore, is hot, its surface colder and heat is still being lost. Interestingly though, if we extrapolate the increase in temperature in deep mines to the Earth's interior, around 15°C/km, it gives us a figure of 95,500°C at the centre, while in actual fact it is 'only' about 6000°C. The Earth's crust is a good blanket and keeps the heat in, just as a duvet separates a hot you from a cold outside (at least in the Highlands). There is a hot Earth below the crust. If you want to test this, insists one book, try putting a pencil-shaped piece of slate into the kitchen gas flame and waiting until it is too hot to hold; you will wait a long time. Despite this, the Earth *is* cooling and what is important is how it does this. All the planets are cooling; what gives them their individual character and determines if they are pock-marked or not, is how they do it.

The rule in geology is that we should look at the present to understand the past ('the present is the key to the past'), a concept introduced in Chapter One and referred to several times since. However, applying this concept blindly to extreme geological lengths of time is being far too hopeful; the butterfly is hardly recognizable from the caterpillar. However, to explore the cooling history of the Earth we need at least to look at heat loss today (to follow the philosophy) but then, considering its hot start and realizing that there have been fundamental changes over time, try to discover whether the Earth is still in its active, 'hot' years, in its cool middle age, or in its stiff and painful old age. To begin we shall look at heat loss today and then consult the Lewisian for the deeper past, maybe elsewhere than at Cape Wrath.

It may have originally been a homogeneous dust ball 4.56Ga ago, but today the Earth is layered like an onion. We know this from studying earthquakes. An earthquake explosion is 'heard' around the globe, meaning that it can be detected on sensitive, sound-measuring seismographs, which is exactly what happened with the devastating Sumatra event at the end of 2004. For the past 50 years or so there has existed a global array of seismic

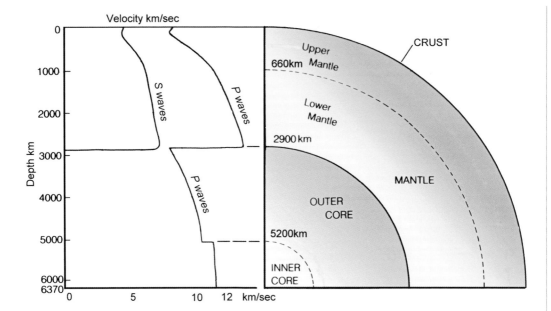

Figure 6.4
The Earth is layered internally, the structure being revealed by earthquake wave velocities. The outer solid layer, the crust, is extremely thin, being only 10–30km thick.

listening stations, originally established for America to listen to Russia testing nuclear bombs and for Russia to listen to America doing the same thing. These stations, although they still monitor for man-made explosions, are now principally used for scientific purposes, which is what allows us the onion model of the Earth. Perhaps it would actually be better to call it a melon model. A melon's layers change from the outside inwards, from skin to flesh to pips; an onion does not change. The layers of the Earth change inwards, downwards, from skin or crust to core; hence the melon model. In an earthquake, although some sound travels along the surface, most energy spreads out and passes right through the Earth itself, but with different speeds (velocities) in the different layers. From the different arrival times of the earthquake shock at the listening stations, the melon model can be built. Think of it as taking an x-ray of the Earth's interior. According to these observations, below a very thin crust (examined later), the Earth has an upper and lower mantle of solid silicates (silicon oxides), an outer core of liquid iron and a solid iron-nickel alloy central inner core. The mantle, the outer layer with a base at 2900km, about halfway to the centre, is both a solid and a very viscous liquid and, like silly putty, what you call it depends on how you treat it. On a day-to-day basis it is solid and brittle – hence the shock of earthquakes when it breaks – but over geological time periods the mantle flows very, very slowly at rates of 2–15cm per year. That is equivalent to taking one million years to move 20–150km, the sort of speed you feel your car is going some mornings in the commuter jam. Modern geophysical analysis can even create pictures of these slow-moving mantle currents, using a technique called tomography. Tomographic pictures actually show where sound travels faster or slower than normal, because in hot, light, rising mantle currents sound will travel more slowly, while in cold, dense, sinking currents it will travel faster.

Tomographic pictures show that the mantle is churning with convection currents, the same as in the morning porridge boiling away in the saucepan. Some of the currents seem to move only through the upper mantle, others rise the full 2900km from the limit of the liquid outer core, although to complete the full cycle from core–mantle boundary to the surface and back may take perhaps 300Ma, a motion so slow that a tomographic map is actually a time library of past, present and future activity. The convection currents transfer heat from the centre of the Earth to the crust beneath our feet and, even though the transfer is very slow, they are cooling the Earth. Just as the surface of the morning porridge moves about as it boils, so does the Earth's surface, its skin (geologically, crust), following the convection currents below. In human terms we see these convection currents at the surface as moving continents, forming mountains, disappearing oceans and earthquakes. It is convection currents in the mantle that make the Earth's skin move about, the driving force being ultimately the Earth trying to cool down. Although convection currents are deep within the

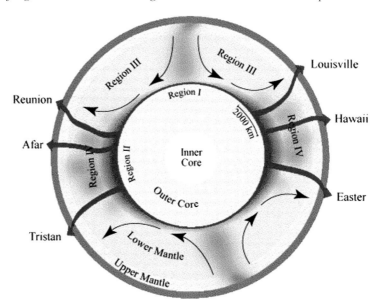

Figure 6.5
Convection currents, which act as a cooling mechanism, churn away in the Earth's mantle moving at speeds of 2–15cm a year. When they reach the surface the currents cause continental drift. Near-equatorial cross-section through the Earth. Red lines are mantle plumes. (From Gonnermann et al., 2004, Reproduced with permission of ELSEVIER BV.)

Earth, we can only study their direct effects in the rocks of the crust, realizing that with a radius of nearly 6378km (4000 miles) this crust, often only 5–10km thick under the oceans but up to 80km under the Tibetan Himalayas, really is just a very thin skin.

If men from Mars, always considered to be green for some reason (what colour do they think we are, pink, black or yellow?), studied the Earth out of curiosity like we do their planet, then one of the first things to strike them would be the remarkable jigsaw fit between the bits of brown land they see showing through the blanket of blue seas. They would surely make little jigsaw models to see how best the bits fit together. Which, of course, they do very well. The German meteorologist Alfred Wegener famously noticed this in 1910, especially the fit of Africa and South America, and promptly proposed that these continents had indeed originally fitted together and hen split apart. He boldly but logically proposed continental drift, only to see *himself* set adrift, both he and his theory being ridiculed by the geological establishment. This is of course history; now we can even measure how fast the continents are moving and in which direction, using Global Positioning System (GPS) stations: slowly and in every way. The mechanism is now called plate tectonics, as it has been found that the continents are actually carried around on 'plates' like bits of left-over dinner. The plates are the hardened crust of the moving convection currents beneath. To a space-based observer, however, it is the continents, the left-overs, that move. It truly is 'continental drift'. The term is disliked by specialists because it is associated with a rejected theory that was in fact correct, establishment science being shown to be wrong yet again. So, looking down from Mars, what do we see?

As the continents slowly drift about at 2–15cm a year, carried by the convection currents, there is nothing to see in their wake. But, as a carpet pushed over a polished floor ruckles when it encounters an obstacle, the front edges of continents form lines of rising folds. On Earth, these are seen as mountain belts that rim the leading edges of the moving continents. Sometimes mountain chains like the Andes form overlooking oceans; sometimes two carpets meet head-on and both ruckle, to form lines of mountains like the Himalayas between India and China. From space, this is all we see. On Earth, when we examine the mantle currents that carry the continents, we find that where they move away from each other, long straight cracks appear, into which new, molten magma wells up from below. These cracks normally appear deep below

Figure 6.6
The 'fit' of the coasts of Africa and South America was noticed before the theory of Plate Tectonics. It was noticed by A. Sneider in 1858 and by Alfred Wegener in the 1920s.

the oceans. If they normally occurred on land where we could actually see them, then perhaps Wegener's theory would have been accepted more easily instead of being neglected for more than 50 years. Now that we know what to look for, of course, we find that there actually are places where we can see the cracks and even walk across them, Iceland being one such place, as we found in Chapter 4. The mid-Atlantic Ridge, actually the mid-Atlantic split, passes right through the middle of the island, the northwest of the country moving away from the southeast at a rate of a few centimetres a year. The volcanoes and lava flows are the symptoms of a scar in the Earth's skin that will not heal.

To be more geological and scientific about all this, the back edge of a moving plate is called a divergent margin. At a divergent margin new crust is formed as the plates (actually the convection currents) move apart, the rear of the plate being simply a split that is continuously filled with magma from below. The front edge of the plate, where mountain belts form, is called a convergent margin. Along a convergent margin, crustal plates (again the currents) descend into the mantle creating deep ocean trenches as they go. The process is called subduction. The continents, however, are low in density and are scraped off (left-over dinner again) to remain at the surface, though a bit crumpled and perhaps containing a few thin bits of plate as well. Plates are forever being recycled, created anew along divergent margins, remelted and absorbed along convergent margins. Where plates rub against each other there are long, linear faults like the San Andreas, along which there are frequent earthquakes. These are transform faults. Through all this convection current activity, the continents (the left-overs), stay on the surface to make up the broken and ruckled bits of the jigsaw that we now try to fit together. The result is that the continents are made up of all sorts of bits and pieces, some that have been around a long time, others that are quite new. The oldest bits, as we know, are 4.0Ga old, almost (but not) as old as the Earth itself. The oldest oceanic plates are 'only' 200Ma old (Lower Jurassic) and most are less than 60Ma old (Tertiary). The theory of continental drift makes us think we understand the present Earth's surface. We live on top of a dog's dinner, on bits of continental left-overs.

What we see, then, on the surface of Earth today, losing heat by mantle convection, is a skin heaving, moving, cracking, wrinkling and being selectively recycled. It explains the differences in age of the rocks now making up that skin and how bits of very old continent like the Lewisian are still preserved as left-overs carried about on the moving plates, which themselves are constantly recycled. But is that all? The second law of thermodynamics tells us that this story cannot go on unchanged, there being no such thing as a steady-state thermal regime; a fire always burns out. Dr James Hutton's coal layer heating the Earth was criticized for just this reason: how could it go on burning for ever? And this is where the Lewisian gneiss becomes interesting; it contains clues about these changes and about the Earth's past behaviour. As we said, this chapter is about forcing a geological confession from the rocks, a confession about progressive loss of heat. So let's go back now and look at rocks. Let's look at the Lewisian.

We started out looking at the Lewisian at Cape Wrath, but perhaps we can actually avoid going back there this time and visit instead the area around Scourie and Loch Laxford, a few tens of miles to the south, a place more welcoming than the former and beautiful in a bleak, minimalist kind of way, the white, naked rocks attractive to those who study them. Even here, though, the Lewisian has a reputation, and to quote from that famous Peach and Horne memoir of 1907:

Figure 6.7
Diamonds are often very old, even over 3.0Ga, and are mainly found in Archaean rocks like the Lewisian that make up the cores of the continents ((photo © MarcelClemens, Shutterstock).

Along the western seaboard of the counties of Sutherland and Ross, the Lewisian or fundamental gneiss forms an interrupted belt stretching from Cape Wrath to Loch Torridon, and thence to the islands of Rona and Raasay.

They then add, somewhat emotionally:

Throughout this belt of country, bare rounded domes and ridges of rock with intervening hollows, follow each other in endless succession, forming a singularly sterile tract, where the naked rock is but little concealed under superficial deposits, and where the surface is dotted over with innumerable lakes and tarns.

The area clearly imposed emotional feelings on them, they certainly must have tramped it for many hours and in this emotional outburst, surrounded as it is by dry Victorian geological statements, their feelings come through like rocks out of a Lewisian bog. It is indeed a 'sterile tract' but all the more memorable and haunting for it. And even Peach and Horne do not mention *Culicoides impunctatus*, the midge; nobody does, but it is here in its millions, billions, trillions...

Many academic theses have been written about the Lewisian Gneiss. Many analyses have been published and still continue to be. These rocks fascinate geologists. Besides being old they are complicated and difficult, human enough characteristics, and they are remote. But there must be more than this; any obsession has both obvious and hidden aspects. It is as though these rocks somehow contain the runes of life itself and provoke the sort of feelings one gets from reading *The Lord of the Rings*. This is pure fantasy, of course. There is certainly an essence to northwest Scotland, a presence without there being one, a 'truth' with there only being imagination. All there really is, is rock. So, with hammer and geological map in hand, this is what we shall look at – the rocks – although with so many experts working with the same information, interpreting what the experts mean is as difficult as interpreting the rocks themselves. We will choose one story and let the others languish, which may not be scientific but is certainly practical, and try to find in the actual rocks an ancient-Earth story.

A little north of the quiet coastal village of Scourie and nearly three kilometres eastwards into a true piece of low, rolling, sterile Lewisian tract, is the small Gorm Loch: really a lochan. You need patience, good boots, fine weather, a map and geological knowledge to find it; patience because you will walk around lochans, by-pass hills, avoid cliffs but never in a direct line; good boots because you will walk on bare rock, deep heather and in soft, wet bog; fine weather because it is far more agreeable; a map because it is easy to get lost; geological knowledge because following the 'grain' of the rock makes finding your way to a precise point, when you are not able to walk in a straight line, actually possible. Even a short, three-kilometre trek in

Figure 6.8
In the words of Peach and Horne, a typical 'sterile tract' of 'bare rounded domes and ridges of rock with intervening hollows'. The Lewisian Gneiss near Stoer.

this terrain makes Peach and Horne's description of it seem restrained. Imagine: during 14 years they had to examine all this area, bit by bit, bog by bog, endless exposure by endless exposure, with heavy overcoats and hob-nailed boots, perhaps ponies, but certainly only a horse and carriage for transport to the hotel at the end of the day. No petrol engines. With Gore-tex and light-weight boots, the short walk is perhaps not all pleasure, but well worthwhile and, in the right weather, quite breathtaking. As you cross the bog, trembling arctic cotton warning you of its softness, the sun glints on the ripples of even the shallowest lochan, shining on the wild water lily pads and lighting the deep yellow hearts of the white flowers like floating Chinese candles. The peak of black Ben Stack and the grey ramp of Arkell, mountains of the Moine Thrust, appear in the eastern distance, rising above the nobbly, bleached rock outcrops as a ridge is scaled. From the tops of the highest crests, looking behind you back to the west, the land disappears somewhere into the sea of the Minches, it is hard to see exactly where.

Between Gorm Loch and Clar Loch Mor, which is a kilometre to the south, a careful, detailed geological map has been made. This means that every bare piece of rock has been measured, sampled, described and registered on a large-scale map, to provide very detailed patches of information. They are separated by lots of white spaces where there is blind bog. The final geological map that everyone sees is drawn as if all were bare rock, and is completed by making geological guesses at the boggy bits between the exposures, like laying out a jigsaw puzzle with more than half the pieces missing and then completing the picture. In truth, for the geologist, there are more than enough bare bits of rock to make an excellent geological map, and it shows beautifully that there is order in the Lewisian, even in the middle of such truly ancient complexity. At first view, the area seems

Figure 6.9
The detailed
geological map
between Gorm Loch
and Clar Loch Mor in
the Lewisian 3km east
of Scourie. It shows
an isoclinal synform
of Early Scourian basic
and ultrabasic igneous
sheets in the steeply
dipping grey gneisses.
Violet = basic, Red
= ultrabasic, Black =
Scourie dolerite dykes.
White = older, grey
gneiss. (After Beach,
1978.)

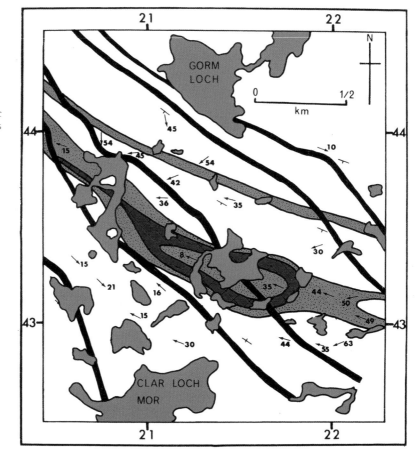

to consist only of fairly regular rock layers, up to a metre or so thick, of the now familiar grey, weathered gneiss, dipping steeply at 70° or more to the southwest. A closer look, however, shows that the gneiss has a fine banding on a centimetre scale, in greys, whites and dark colours, beautifully picked out by a combination of ice smoothing and several thousand years of rain and weather. The climate of Scotland is very good at this. But be curious, follow a single layer; it will not be continuous. Frequently it will turn back on itself in a tight hairpin to make the kind of shapes that form in a heavy, hanging curtain, or too much toothpaste squeezed onto a toothbrush, a squashed 'C' and geologically called an isoclinal fold. The rock appears to be just layers because it has been so compressed that all the irregularities have been squashed flat. When a car leaves the scrap yard, no matter what model or shape it was when it went in, it comes out as the same-sized compressed cube. These Lewisian rocks have been so squeezed that they have all come out in the same shaped isoclinal folds, from a tiny scale of less than a centimetre, to a scale of metres, or even kilometres, as we shall see in the geological map. Such structures were in the cliff face at Cape Wrath and, indeed, can be recognized all over the Lewisian outcrop. But to the geological map.

In the middle of the mapped area centred around an unnamed lochan are dark (mafic) rocks, geologically basic and ultrabasic types. The ultrabasic rocks are still finely banded like the grey gneisses, but all in dark minerals like olivine, hornblende and pyroxene, some bands standing out from others. Along with these dark rocks there are soft, brown schists, a rock with platy micas, some of which here have huge, dark red garnets (a much underrated gem). By tracing all these distinctive rock layers which, remember, are standing on end and nearly vertical, it is possible to map out an isoclinal fold a kilometre long, geologically termed a synform at this scale since the closed hinge points downwards. The geological map, a picture of the surface of the Earth's skin, is of course a horizontal cut through the geological structure and makes the fold look like a knot in a plank of wood. The vertical profile of this structure, though, is the same squashed 'C' as seen in the smallest scale isoclinal folds just described. Most important, however,

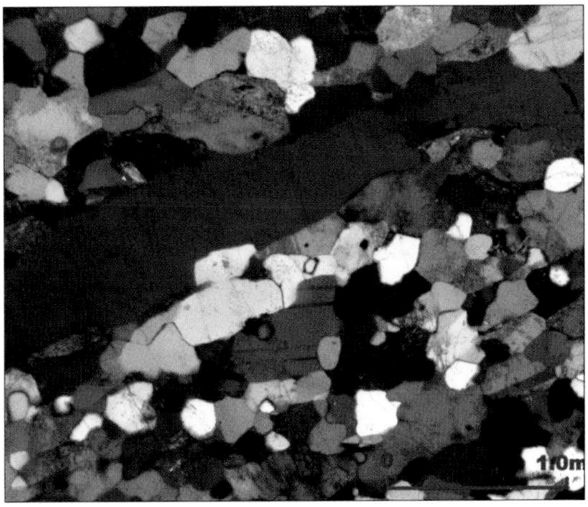

Figure 6.10
3Ga grey gneiss seen in a thin, transparent slice of rock (thin section). The bright aligned crystals are micas, the irregular grey masses are quartz crystals. Bar is 1mm.

is the fact that such large pieces of dark, basic and ultrabasic rocks and related schists are still together, despite having been completely interfolded and interleaved with the surrounding grey gneisses. All these rocks have been very heavily metamorphosed, that is, recrystallized without actually melting; they were too deep in the Earth for that. To form the minerals found at this location needed extreme heat, over 1000°C, and pressures of over 15kb (kilobars), which translates into conditions 35km below the surface, effectively the bottom of the crust (skin). The geological term 'granulite facies' is given to these severe conditions, to which geochronological dating gives an age of 2.7Ga and geologists give the name of 'Scourian' metamorphism.

What turns out to be important about the Gorm Loch map is that it shows that the light-coloured rocks, the dark-coloured mafic (ultrabasic) rocks and the platy schists are all still individually visible. The dark rocks are thought to have originally been early oceanic crust and the platy schists the ocean bottom sediments deposited on them. The light-coloured rocks are considered to have been igneous rocks from an early type continental crust, with an age of 2.9–3.0Ga (200–300Ma older than the Scourian metamorphism that changed them). This means that the Lewisian Gneiss is made up of primitive Archaean continental crust that was forcibly squeezed and intimately interfolded with Achaean oceanic crust, 35km below the surface, 2.7Ga ago. These conditions are interpreted as indicating something like a subduction zone, that is, an Archaean convergent margin. A real terrestrial *mille feuille* if you like (a thousand layers, one of the best French pastries to have with morning coffee). However, this *mille feuille* is special.

Today, as we have described, continental crust is lighter than oceanic crust, and when the two collide at a convergent margin, the continent rides over the oceanic plate like a surfer does a wave (or the continent is scraped off, if you prefer the left-overs image). In the Archaean, when a collision occurred it was like two surfers crashing head on, the two getting totally tangled up with each other. The oceanic crust and the continental type crust were hotter and showed less density difference then than now. The geological map of Gorm Loch shows us the intimate, working details of a 2.7Ga Archaean collision that could only have happened at a higher temperature and higher velocity than today. We are forensically examining the twisted tangle of the beginnings of continent creation. Such collisions have not occurred since the Archaean.

There is something wrong, though. Look at the geological map (Figure 6.9). Clearly standing out are very simple, more or less linear features like roads, 30–40m wide, cutting across the 'grain', and complex folds in the gneisses like a vandalizing chainsaw cut through a polished wood table. They are quite out of place. These are the lines of the so-called Scourie dykes, and because the rocks that form them are eroded more easily, they do not stand out as rock features but appear as long, straight-sided, shallow gullies, filled with light green bog moss and grass. Several dykes cross the area and can be followed for over 4–5km, heading southeast or northwest, without an end being found. In places, enough rock sticks through the bog and grass to show that the dykes are formed of dark, massive dolerite, but they are difficult to study here. A similar dyke, perfectly exposed and easy to study, is found near Scourie, the coastal village that has involuntarily given its name to a unique geological event of dyke intrusion that took place 2.4 billion years ago. Scourie in recent years has awoken to this heritage, and at the centre of the Geopark is planning a visitor exhibition.

A short walk from the Scourie village jetty northwards over the cliff tops leads to a location that to specialists of the Lewisian is like Harvey Nichols to a shopper; a must-visit place. Every rock along the coast here seems to have its own name, however inappropriate. So giving another, geological name is simply keeping up a tradition shared with the Picts, the Vikings and the Gaels. At a place called Creag a' Mhàll by the Gaels, a steep scramble down to the bottom of the cliff takes you onto a geologically named Scourie dyke, *the* Scourie dyke. It shows all the key characters, was featured by Peach and Horne, and is visited by all serious students of the Lewisian. This is the Mecca of Lewisian geology. This dyke, as any dyke by definition, is a vertically sided igneous injection, like a wall, which here is about 40m wide. Because it is more easily eroded than the surrounding rock, the dyke forms a gully and a narrow platform running in a straight line right along the cliff bottom at sea level, back towards Scourie, creating a small bay half a mile away where it makes landfall. (The bay, of course has its own Gaelic name: Poll Eorna.) It then continues inland heading south-eastwards for many kilometres. Meanwhile, down on the rocks at Creag a' Mhàll, if you break off some of the dyke it is a black, crystalline dolerite, a rock in the basalt family. To the eye it is without distinctive, identifiable minerals, because they are too small to be seen. Under the microscope, however, in thin section, small lath-shaped crystals of plagioclase feldspar can be seen, and these are surrounded by dark crystals of clinopyroxene, which is what makes it a dolerite: nothing spectacular. But if you look at a sample taken from the margin of the dyke, even with the microscope at 100 times magnification, it is difficult to see any crystals at all; they are too small. This is evidence that the edges of the dyke were chilled so rapidly as the dyke was injected that normal-sized crystals could not form. Only in the centre of the dyke, where heat was retained, could larger crystals grow. Clearly, the dyke rock was hotter than the gneiss, a prerequisite for the chilled margins, and also as it is denser than gneiss when at equal temperature; it had to have been hot to be injected. By this time the invaded gneiss must have cooled right down following the preceding Scourian metamorphism.

So why is this such an important location? If you look at the rocks in the cliff and to the side of the dyke, they are all squashed, folded and completely mixed from the 2.7Ga crustal collision (Scourie

Figure 6.11

Geological map of the
Lewisian basement
between Quinag,
Suilven and Canisp
showing the many
Scourie dykes crossing
the area (after Cadell,
1896).

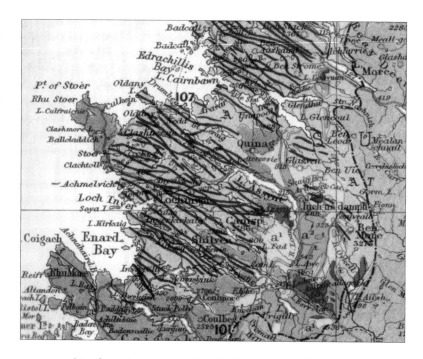

metamorphism) and yet here, in all the rock confusion, is a quite
simple, vertical, igneous dyke. It has sharp, clear, chilled contacts
and is still in the vertical position in which it was originally injected
billions of years ago. Geologically it is a gem. It means that the ocean–
continent collision can be perfectly dated; it had to have happened
well before the dyke was injected. It is a clue so simple that, for
once, there is no argument, even between specialists. In addition,
the basaltic composition of the dark-coloured dyke itself is distinct
when compared to the surrounding light-coloured, grey, quartz-
rich gneisses. This and nearby dykes have ages of 2.4Ga, that is,
300Ma younger than the ocean–continent collision and main gneiss
metamorphism. In Earth terms, the dyke injection is a completely
new event, an opening up and splitting of the crust rather than a
compression and rock creation as during collision. A glance at the
geological map shows that in the region between Loch Laxford and
Loch Maree near Ullapool 65km to the south, there are many Scourie
dykes like the one we are standing on, all crossing the Lewisian
outcrop in more or less simple, parallel, northwest–southeast and east–
west trending lines. The collective name for them is the Scourie dyke
swarm and they represent a big, 2.4Ga geological event (although
some dykes could be slightly younger). That the simple trace of most
of the dykes can be followed along miles, in this otherwise hopelessly
complicated rock, is quite remarkable. It is like driving your car down

the empty bus-only lane in central Edinburgh at 8.45 in the morning while the rest of the traffic is jammed; it makes everything seem so simple!

However, it is not the same everywhere. In the Lewisian south of Ullapool, about 56km south of Scourie, dykes can be frequent but their traces are not simple and they have been bent, broken, faulted and modified, although still recognizable as dykes. This is the case at Gruinard Bay, the next inlet south of Loch Broom (by Ullapool). Dykes are common and although they are bent and broken, they are sufficiently identifiable to be seen heading towards Gruinard Island, infamous for having had anthrax spread over it as an experiment in 1942 during the Second World War (it didn't actually stop the locals using it, which might explain a thing or two). The island was only officially declared decontaminated in 1990 (48 years later) after having been sprayed with formaldehyde (a poison! – like whisky) and seawater. A different effect is seen near Scourie. For the few miles north of the village and as far as Loch Laxford, just beyond the bird island of Handa (no anthrax here, thank goodness), the dykes are simple and similar to the one at Scourie itself, examined previously. But then, along Loch Laxford, there is a tectonic 'front' or abrupt change of structural state, and from here northwards all the way to Cape Wrath there are no clear traces of dykes. In the many road cuts made recently to let the modern road through to Durness (and Cape Wrath), grey gneiss is mixed with chopped up bits of dark, massive rock, which is all that is left of the Scourie dykes; they have been

Figure 6.12
The oldest Saltire. A St Andrew's cross in the dark: 2.4Ga old dolerites of a Scourie dyke, broken and injected during post-Scourian metamorphism. Hammer 28cm long.

tectonically ingested into the gneiss and, like chunks of meat in pet food, are almost unrecognizable.

Mainly from the character of the Scourie dykes, the mainland Lewisian outcrop from Cape Wrath all the way south to Raasay and the Inner Isles, a distance of over 160km, was divided by Peach and Horne into three, geologically distinct areas: north, south and central. Only in the central area, between Loch Laxford and Loch Broom, are the dykes as clear and intact as at Scourie. To the north and south, as we have described, they have been faulted, bent, stretched and broken up, especially to the north. Later workers, basing their arguments on this dyke behaviour, proposed several separate, major geological events: those pre-dyke (the Scourie metamorphism that was mapped at Gorm Loch); the event creating the (Scourie) dykes themselves; and post-dyke (post-Scourie) events that so distorted and changed them (Scourie does seem to have been born to geological greatness). Most recently, it has been suggested that the north, central and south areas are in fact different bits of continental crust (called terranes), tectonically stuck together, like different patches on a quilt. As more workers have examined the Lewisian and more and more geochemical and especially geochronological (geological age) analyses have been made, more major Archaean and early Proterozoic, Earth-forming (tectonic) events have been discovered and dated. Today's specialists now talk of eight or more different bits of continental crust stuck together in nine principal tectonic events spanning 1.8Ga, a huge length of time. Geologists are obsessed with the Lewisian, as mentioned before, so the details of all these terranes and events do not matter, especially as when one expert discussing one particular event states:

> *The metasedimentary gneisses described [...] contain a corundum-kyanite-stauroloite-plagioclase assemblage partly retrogressed to margarite-paragonite-sillimanite-chlorite and are interpreted as restites arising from partial re-melting.*

Which for the average person is about as difficult to understand as the rocks themselves. But after all this, what *do* these eight terranes and nine events at Scourie and Laxford actually mean? What is the significance of the banding, the complicated structures, Scourie dykes, isoclinal folds and unique minerals, in terms of the Earth's geological history? Because this is why we looked at the Lewisian in the first place. Let us outline the essentials of the story, going backwards in time, looking first at the youngest event.

First: the Lewisian has remained untouched for over one billion years. This simple fact speaks much; these rocks were at the surface by 1.2Ga, and have not been involved in any significant geological activity since then, save a bit of bobbing up and down. They have obviously survived until today. Second: look back further and we see that these rocks were modified deep below the surface but brought persistently upwards to their present position in complex tectonic activity, the eight terranes and nine events we have looked at. Throughout these events, continental crust was being gradually created, squeezed together, metamorphosed, pushed up and added to from below (accreted), while the planet was cooling and the continental crust beginning to grow. And then third: the prime event deep in the past, the creation of the original bits of Lewisian crust, at high temperature and depth, for the crust, at the great age of 3.0Ga. In the correct time sequence these events add up to a long story of planetary cooling: hot, deep, early continent formation; complex tectonics and metamorphism during uplift and cooling; then long-term stability at the centre, and on the surface, of a growing continent. This new continent no longer increases in size by the addition of material from beneath, but by the addition of mountain belts wrapped around the old Lewisian core. The new style of growth is a sign of continued cooling and the beginnings of the modern plate tectonics that were described earlier.

Eventually the Lewisian Gneiss and the first mountain wrappings became the continent of Laurentia, which at one time included bits of present day America, Canada, Greenland and Scotland. The Cambrian

Figure 6.13 Continents have been built up by the addition of mountain belts around Archaean cores for the last 2.5Ga (a(after Holmes 1996)).

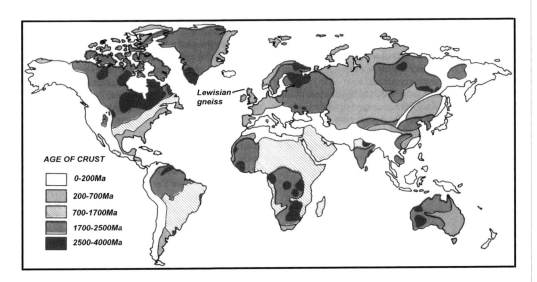

AGE OF CRUST

0-200Ma

200-700Ma

700-1700Ma

1700-2500Ma

2500-4000Ma

Lewisian gneiss

we looked at in Chapter 2 was deposited on and around it. Across the planet, all the continents are like this: Africa, India, Australia, China, South America – an ancient heart wrapped in layers of younger and younger mountain belts. The Lewisian comes from a special time in Earth's history, when continents were forming and events of creation occurred that are not possible today. As the Earth has cooled, so the continents have become bigger and fewer until now, huge plates and long, linear features are what we experience. The hot, energetic Lewisian Earth has lost heat and stiffness has set in. Like all things geological, the cooling and the changes have been dreadfully slow, but equally characteristically, they have been unremittingly persistent. And they will continue. However, having survived for 3.0Ga, the Lewisian will now survive for as long as the Earth.

Tolerating the ridiculous discomfort of Cape Wrath on a bicycle is worth the lovely experience of being able to see a rock with which the Earth was creating continents. When we touch the Lewisian Gneiss we are touching the heart of a young continent, a rock reminder of our planet's beginnings. A venerable fossil – or nothing but geological left-overs, a dog's dinner: it depends on your view.

Further Reading

Books, Pamphlets

Barber, A.J., Beach, A., Park, R.G., Tarney, J. and Stewart, A.D. 1978. *The Lewisian and Torridonian Rocks of North-West Scotland.* Geologists' Association Guide, No 21. pp.99.

Condie Kent C. 1997. *Plate Tectonics and Crustal evolution.* Oxford: Butterworth Heinemann. 4th Edition, pp.282. ISBN 0 7506 3386 7

Condie, Kent C. 2005. *Earth as an Evolving Planetary System.* Cambridge, Massachusetts, USA: Elsevier Academic Press. pp.447. ISBN 0-12-088392-9

Craig, G.Y. (Ed.) 1991. *Geology of Scotland.* Geological Society London. pp.612. ISBN 0-903317-64-8

Johnstone, G.S and Mykura, W. 1989. *British Regional Geology: The Northern Highlands of Scotland.* British Geological Survey, Her Majesty's Stationery Office. 4th Edition, pp.219.

Juteau, T. & Maury, R. 1997. *The Oceanic crust, from Accretion to Mantle Recycling.* Berlin: Springer-Verlag. pp.390. ISBN 1-85233-116-X

Peach, B.N., Horne, J., Gunn, W., Clough, C.T., Hinxman, L.W. and Teall, J.J.H. 1907. *Geological Structure of the North-West Highlands of Scotland.* Memoir of the Geological Survey of Great Britain. HMSO. Glasgow. pp.668.

Van Andel, T.H. 1994. *New Views on an Old Planet.* Cambridge: Cambridge University Press. 2nd Edition, pp.439. ISBN 0 521 44755 0

Wooley, A.R. 1981. Growth and development of the continents. In: Cocks, L.R.M. (Ed.) *The Evolving Earth. British Museum.* Cambridge: Cambridge University Press. p.21–33. ISBN 0 521 28229 2

Scientific Papers

Gonnermann, H.M., Jellinek, A.M., Richards, M.A. and Manga, M. 2004. Modulation of mantle plumes and heat flow at the core mantle boundary by plate-scale flow: results from laboratory experiments. *Earth and Planetary Science Letters*, 226, p.53–67.

Kamber, B.S., Moorbath, S. and Whitehouse, M.J. 2001. The oldest rocks on Earth: time constraints and geological controversies. *In*: Lewis, C.L. & Knell, S.J. (Eds) *The Age of the Earth from 4004 BC to AD 2002.* Geological Society London, Special Publication 190. p.177–203.

Park, R.G. 2002. *The Lewisian Geology of Gairloch, NW Scotland.* Geological Society Memoir No 26. Geological Society London. pp.80.

Park, R. G. 2005. The Lewisian terrane model: a review. *Scottish Journal of Geology*, vol. 41. pp.105–118.

Rollinson, H. 2001. Searching for the earliest record of life on Earth. *Geology Today.* vol. 17, No 5. pp.180–185.

Sutton, J., Watson, J.V., 1951. The pre-Torridonian metamorphic history of Loch Torridon and Scourie areas in the northwest Highlands; its bearing on the chronological classification of the Lewisian. *Quarterly Journal of the Geological Society of London.* 106. pp.241–307.

Chapter 7   HUTTON'S ARSE (1795)

Just another planet?

I n 1795 Dr James Hutton (1726–1797), citizen of Edinburgh, physician, merchant, inventor, farmer, bachelor (?) and founder member of the Royal Society of Edinburgh, published a scientifically seminal and lastingly famous book, *Theory of the Earth* in two volumes. As a person we see him now through the austere picture painted by a young Henry Raeburn (1756–1823) when Hutton was about 60, mostly in dull, presumably intentionally earthy browns and blacks, seated, with his hands primly clasped in his lap and staring without focus as it were into endless geological time. 'We see no vestige of a beginning, no prospect of an end' is the look on his face and are the words with which he most famously concluded the final volume of his book. The image of a serious, pious, reflective eighteenth-century philosopher of the 'Scottish Enlightenment' (late 1700s) is complete.

> *My arse, it is evident, is now a part of much greater consequence* [...] *than my head...* [or] *...lord pity the arse that's clagged to a head that will hunt stones...* [or even] *...my arse to the east and face to the Irish Channel...*

...(when heading for Wales) are the astonishing words of the flesh-and-blood man. Is this the original Dr Hutton and Mr Hyde? Dr Theory-of-the-Earth and Mr Arse? We always thought that Louis Stevenson based his story on Edinburgh's Deacon Brodie but perhaps there were precedents.

The late 1700s are now too far past for us to know the truth, so we must divine it or invent it. It is sure that Hutton had most important and original scientific ideas, and wrote a seminal book on the geology of the Earth, but neither of these make us ponder on the state, importance or orientation of his arse. The picture by Raeburn is awful, certainly in terms of artistic merit, so we do not even know if it represents a likeness. The artist does not seem to have been able to paint wrinkles on a face and certainly could not paint hands or place a sitter on a chair properly. The picture looks even worse in the original (in the Scottish National Portrait Gallery, Edinburgh) than it does in small reproductions. It is not of a 60-year-old man:

> *Hutton geologist, in quakerish raiment, looking altogether trim and narrow, as if he cared more about fossils than young ladies...*

gibed Robert Louis Stevenson (1850–1894) of this portrait, although of course he did not know him personally. I think we have understood the theory but misjudged the man, the 'Father of Geology', but also of an illegitimate son.

This final chapter is not about James Hutton but is inspired by him. It is different from the other chapters in that it is not attached to any particular place in the Highlands. It is not supposed to incite a visit to any particular strath or glen. It is simply an attempt to consider the Earth as just another planet and to put the previous chapters, and geology in general, into context. It has been difficult to write. It is not a 'Theory of the Earth' as Hutton wrote, but in 10 pages, nor an 'Earth' chapter in the *Hitchhiker's Guide to the Galaxy*. It is an attempt to think of the Earth from the outside, not as a local would do, but as an incomer. Can an ant ever imagine its own heap? No! Of course not; so perhaps neither can we, although at least, being human, we can try. Hutton became important to this chapter because of his attitudes. He somehow managed to detach himself from the world on

Figure 7.1
Portrait of James Hutton. This awful picture by Raeburn, painted when Hutton was about 60, is the only one we have of him. We do not even know if it is a likeness (courtesy of the Scottish National Portrait Gallery).

which he lived, and looked upon the Earth as an entire object. He watched his good friend James Watt of Edinburgh (inventor of the steam engine) building his heat machines, and considered the Earth to be something of the same. Everything on the planet worked together, like a heat engine: an Earth engine. He saw through the detail of everyday minutiae to the scale of the entire planet, and from his careful observations of actual sediments and real rocks, he knew that his Earth machine normally worked very, very slowly. However, it is punctuated by catastrophic rapid events. Importantly, these theories were based on what he saw. Hutton liberated himself from Genesis; we must liberate ourselves from Gaia. Huttonian attitudes will help – but we must leave his arse out of it.

As American, Russian, Chinese and occasional European space probes have wandered through the Solar System, brushing planets and discovering and photographing new space objects,

Figure 7.2
The Earth as an isolated planet: a disturbing picture (courtesy of NASA).

we have all been affected. When we first looked at those memorable pictures taken by the Apollo lunar missions, of the Earth as a blue and white globe with brown, recognizably shaped continents, we were shocked. The Earth is perfectly isolated in space – like any other planet. The realization was forced on the human psyche. Scientists felt as if they were looking for the first time at some newly-discovered space object. In truth, we all were; we were at last looking at our Earth as a planet. The revolution in attitude was seminal. David Hume (1711–1776), philosopher of the Scottish Enlightenment and equally a friend of James Hutton, would have been pleased. 'Reason is, and ought to be the slave of passions', he said; and so it was. Now when our probes reach other planets that have a different and strange atmosphere or dramatic surface conditions, we are forced to wonder if we are not the ones that live on an odd planet. Is it us or them; is Earth normal or an exception? Perhaps it is just one more variation amongst many. How does Earth stand as a planet? When children grow up they continually measure their changes in height by marking a line on the bedroom wall. One will spurt, another will stop. This is what this chapter is about: comparing the Earth to other planets, in the past, now and even in the future. Are we growing in the same way as the others or are we different? On a clear Highland night, away from the city, you can stare into the sky and see so many stars – all with their inevitable planets. We are one in billions. How many Earths are there out there? Or are we one *in* a billion?

At the very beginning of the Solar System the terrestrial worlds were all quite similar and hot planets were formed from coagulations of cold space dust, as described in Chapter 6. The planets may all have started the same – there is no class distinction in pre-school playgroup children – but very quickly size and position (distance from the Sun) began to play a role. If we look at our near neighbours, Mars, and because Man has sampled it, especially the Moon, we see something like a hugely speeded-up video of planetary development. By this is meant that we and our neighbouring planets began developing together, effectively began cooling down together, but then, one-by-one, the others became (what we call) fossilized. They reached a critical point not yet reached on Earth. For some reason we have prolonged a stage which the other planets passed through very quickly. The result is that we can examine the fossilized states of the Moon and Mars, as Hutton did the Earth, to understand both how Earth was but also what it will become. These planets, after all, are close to us. There are fundamental differences, of course, but it is the similarities that shock. There is a book entitled *Dead Mars, dying Earth*. Is it and is it? What does that mean anyway, a dead planet?

The 382 kilograms of Moon rock brought back by the Apollo missions between 1969 and 1973 are the most valuable 382kg of any material ever found, but only to those who think so. Art objects are like this, valuable to those who appreciate them. To any scientist, bits of Moon rock do not mean money; it is impossible to put a price on them, as they are so rare and indeed irreplaceable. Their value is in their use in science. In 2002 a locked safe containing 113g of the Moon (the weight of a heavy letter), a sample from every one of the Apollo missions, was stolen from Johnson Space Centre in Houston and the rock put on the market for $8000 a gram (£6000), a small value of nearly a million dollars (half a million pounds sterling, depending on the state of the exchange rate!). A vigilant Belgian mineralogist, who had replied to an internet advert selling the rocks, helped the FBI (it was actually more than help) to arrest

Figure 7.3
Sample 72415 in the
Houston Moon-rock
collection has an age
of 4.45Ga, the oldest
piece of rock so far
found on a satellite
orbiting a planet
(courtesy of NASA and
Meyer, 2004).

four people before any of the valuable grams disappeared. It seems it is actually an offence in America to possess Moon rock: there is no EU directive (Brussels missed this one!).

One of the remarkable aspects of Moon geology is the way in which samples from the now carefully looked after 382kg have been analysed and used to reconstruct the complete lunar planetary history. Rare Earth Elements (REE), described in Chapter 4, play a key role in this. Just to recap, when a rock is enriched in REE it indicates repeated partial remelting, the amount of enrichment indicating the number of times it has been remelted to a magma. Whisky is only made after repeated distillations; at first there is just water and a few raw ingredients. The more a lava is remelted, the richer it becomes in REE. Mineralogically, Moon rocks are basaltic lavas, meaning they consist of pyroxene, feldspar and some olivine; there are no granites or quartz-rich lavas, and so they are primitive (i.e. original – not yet whisky). They are all old, with ages between 3.2Ga and 3.8Ga, although one sample, number 72415 in the NASA collection, consisting mainly of olivine (a dunite), has an age of 4.45Ga, the oldest rock so far found on any satellite orbiting a planet. The lunar basalts are not quite like Earth basalts because there is no water or free oxygen, nor ever was, on the Moon. They are described as having a Moon 'flavour', although their basic mineralogy is not unusual. They are unusual, however, in the REE they contain. First of all, the REE show an overall ten- to one hundred-fold enrichment, that is, in reference to the original material that made up the Moon. It indicates that they are the result of quite a few episodes of partial remelting and recrystallization, especially those basalts with 50 to 100 times the REE norm. This is not unusual and happens on Earth as basalts get recycled. However, there is one difference that marks out the REE patterns of all these Moon rocks from REE patterns on Earth; there is no enrichment in europium. Europium is a rare earth element towards the heavier end of the series, and having a comparatively small ionic radius is therefore less incompatible. The lack of enrichment in just this one element was a complete mystery when it was discovered, but the reason for this oddity has now been identified and the explanation is remarkable. More often than science would like, discoveries depend on luck; so it was in this case.

Figure 7.4
Harrison Schmidt,
geologist, on the
last mission to the
Moon in 1973, Apollo
17. The study of
Moon rock samples
has contributed
significantly in
understanding the
Earth (courtesy of
NASA).

7. HUTTON'S ARSE (1795)

One sample brought back by the Apollo astronauts from the lunar highlands was of the Moon's primitive crust, and it shows why europium is absent in all the younger (although still very old) remelted lavas. This sample is an anorthosite; that is, it consists principally of light-coloured plagioclase (specifically the calcic plagioclase, anorthosite). Europium has just the right ionic size and charge, under the Moon conditions of no water and no oxygen, to fit into the plagioclase crystal lattice. The plagioclase crystals in this primitive rock are therefore very rich in europium. However, that the element is rare in all but the most primitive samples has a remarkable significance. It means that early on in the Moon's development, there was a moment when all the europium was able to be sponged up by plagioclase crystals. The only way this could have occurred is in a well-mixed, molten magma, which under Moon conditions means a hot, bubbling, global, molten ocean 500 kilometres deep. From space, this Moon glows deep red day and night, like molten gold. As the plagioclase crystals formed, being light, they would have floated to the surface of the magma ocean, taking their europium with them. It is a sample of this early, europium-rich magma scum that was by chance sampled by Apollo. Calculations on the extraction of the element suggest that the bubbling magma ocean had to exist for 200 million years, from the Moon's origin at 4.47Ga to approximately 4.3Ga, in order for it to have been sponged up before the entire ocean finally solidified. Because these early crystals floated over the magma, they were effectively extracted from the melt, and the mineral phases that crystallized later, more slowly and deeper inside the Moon, were left without europium. When these later phases remelted and flowed out onto the Moon's surface, to be collected eventually by the Apollo astronauts, the tell-tale lack of europium was locked into their REE patterns.

This may all be very interesting in terms of the Moon's development, but the real significance is wider; if a molten magma ocean lasted for 200 million years on the Moon, then why not on Earth? The Earth is bigger than the Moon, so the ocean would have been deeper, perhaps 1000km, and have lasted longer, perhaps 500Ma. Where is the evidence on Earth? Can we find any terrestrial anorthosite scum? Apparently not. At least not yet and, according to the specialists, we are not ever likely to, either. Because the Earth's magma ocean was originally much deeper and therefore at higher temperatures and pressures than those on the Moon, heavy garnet crystals would have formed first, rather than light plagioclase, and these first minerals would have sunk to the bottom and disappeared rather than floated to the top. Is this the reason why the oldest rock we have so far found on the Earth's surface (Chapter 6) is no older than 4.0Ga, the Acasta Gneiss? The Earth's age is 4.56Ga. The magma ocean must have formed some sort of crust; where is the solid rock of the first 560Ma? There are those tiny zircons (Chapter 6); do they count? Important though they are, a few millimetres are hardly evidence on which to evaluate 560 million years of planetary history, and after all, size does count. We need a new approach.

So far there is only indirect evidence of those first 560Ma: the Earth's first half billion years. The fact that the Earth is made up of layers indicates that an early molten stage indeed existed. Just as the Moon was able to extract all its europium, the Earth separated into the melon model (Chapter 6) of core, mantle and (presumably crust. Magnetism has been recorded in rocks at least as old as 3.5Ga, which indicates that the inner core and outer core had fully differentiated by then to create a strong magnetic field. Indeed, modelling suggests that the crust was formed by about 4.5Ga. So the Earth, like the Moon, had a molten stage. However, because of its size, the crust was not the

Figure 7.5
Parts of tiny zircon crystals are the oldest dated solid material from the Earth's crust; 'Old Boy' (4.2–4.4Ga) from the Jack Hills, Western Australia (from Peck et al., 2001, Wilde et al., 2001, courtesy of University of Wisconsin, Department of Geology).

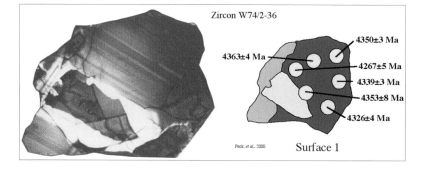

outermost layer. That was (and still is) the atmosphere. This is where we find at least a signal from our earliest planet, not in the solid rocks.

Complicated equipment, all wires, tubes and hissing magnetic switches, that hums away in the occasional university basement, is there to measure the tiny values of rare gas isotope ratios in the present atmosphere. These delicate measurements allow the age of the atmosphere to be determined. This value is 4.4Ga, so far the oldest dated bit of the Earth. It may be surprising for geologists to be dealing with gases and not rocks, but these measured gases originally came from inside the Earth, pouring out of volcanoes, vents and cracks (as they do today), in other words from the degassing of the mantle. Mixed with the gases originally was a great deal of steam, water vapour that condensed to form the oceans. We have already seen what an important role volcanoes played in melting the Snowball Earth (Chapter One). Even before this, though, both air and ocean came from degassing. Is this a guess or do we actually know? We think we know. Of the argon present in our atmosphere now, 99.6% is ^{40}Ar, an isotope only created from the natural decay of the radioactive isotope of potassium, ^{40}K. Common potassium is mainly the isotope ^{39}K and contains only tiny quantities of the radio-isotope ^{40}K. The slow decay of the radio-isotope takes place in solid rock, and the minute quantities of argon 40 gas produced come into the atmosphere only very, very slowly (half-life 1.3Ga), seeping out of the rocks and degassing through vents and volcanoes. So this is where all the argon came from. Since argon originated from volcanoes and vents, then so did the other gases and liquids of the atmosphere, or so the logical argument goes. In fact it seems that from the mixture of elements in the present air we breathe, degassing created a first atmosphere of mainly nitrogen, considerable carbon dioxide, carbon monoxide, methane, the rare gases and a global ocean of liquid water, as early as 4.4Ga. This is only 160Ma or so after planetary accretion. But we have no solid evidence of what was happening to the planet itself. The earliest rocks found on Earth have an age of 4.0Ga, 400Ma years younger (Chapter 6). Despite the existence of an atmosphere and ocean from about 4.4Ga onwards, no rocks from this earlier time have yet been found.

The surfaces of Mercury, Mars and the Moon have not been altered [...] and display the state that was obliterated long ago on Earth...

... said Tjeerd van Andel, doyen of Cambridge University (UK) geology. Was he right about this: has it been obliterated, and was this the state on Earth? Let's look at the early moments on the Moon and Mars to see if these were also the Earth's early moments.

We know that as the first crusts were forming on all the planets, they were being bombarded by an incessant rain of meteors, the final act in planetary accretion. The evidence is in the very old 'surfaces of Mercury, Mars and the Moon [that] have not been altered...'. For example, cratering intensity has been accurately measured and dated on the Moon. Not only was bombardment more frequent just after the Moon's formation (4.47Ga), but the meteors were bigger. Bombardment peaked around 3.9Ga and thereafter size and number diminished through time. The mountains ringing the biggest meteorite craters on the Moon have been much scarred by subsequent impacts and so can be dated as the oldest. During this early time there was an interplay between crust formation and then destruction by meteor impact, rather like the thin ice that forms on our local lochans early in the year, only to be broken by stone-throwing youngsters coming home from school. Where the ice is broken, water seeps through from below; where an early crust was broken, lava flowed through from the molten magma ocean below. On the Moon, these lavas form the flat *maria,* the dark patches we see from Earth, and dated from Apollo samples to at least 3.8Ga; the solid crust itself, as we saw earlier, is dated at 4.45Ga. The surface of the Moon that we can see today is almost as old as the planet itself. Mars is a lot more complicated.

Across the southern two-thirds of the planet are the Martian Highlands. They contain many remarkable features: a 4500km long, up to 11km deep canyon; huge volcanoes 20–27km high; meandering, water-eroded channels and traces of glaciers. This is certainly where the tourists of the future will go when interplanetary holidays begin. You can already see the adverts. But what is important to our present quest is that the Martian Highlands are heavily cratered, and hence old (in Martian stratigraphy, Noachian; i.e 4.5-3.8Ga) and contain the scars of

Figure 7.6
Mars is very lop-sided. It has one large volcano 25km high, one huge, linear tectonic split 2000km long, and two very large meteorite crater scars. It is only one-quarter the volume of the Earth (courtesy of NASA).

very early, very large meteor impacts. Hellas, the largest crater depression, is the most remarkable on maps, 2000km in diameter (the distance from Rome to Edinburgh), and 9km deep; then comes Isidis, 1900km in diameter, Argyre 1200km, and so on. That these craters can still be clearly seen means that a solid crust has been in existence for almost all of the planet's history, 4.4Ga or more, although the flat crater floors indicate that these early impacts freed the molten lava from below the crust that they broke open. Despite all subsequent effects – further impacts, volcanism, probable water, wind and simply time – these early features have easily survived. It is therefore true for both Mars and the Moon that their earliest surfaces are still part of the planet's surface today, as van Andel says? And on Earth? 'Different', say the specialists, 'because of two things, continental drift and water.' Together they keep the oceans young, less than 200Ma old, as we already know. We will never find rocks older than the Acasta Gneiss (4.0Ga) or the Isua Gneiss (3.9Ga) – so they say. All vestiges of the ancient past, as van Andel says, have been 'obliterated long ago'. Yet on the Moon and Mars the first crusts are still visible, and 400–500Ma older than even the smallest pieces of early crust found on Earth.

This is where we need to call on James Hutton to give us some guidance. His famous visit to Siccar Point just outside Edinburgh, in 1794, gives us an insight into his surprising reasoning. Siccar Point, as every geologist knows, is an unconformity between the Devonian and the Silurian: for Hutton it was not. For him it was a difference in geometry and induration between several generations of rock and the present-day sediments on the beach. What he called '*schistus*' (the Silurian) was recrystallized and in beds that were standing on end. His 'red muds and sandstones' (the Devonian) were horizontal, but were very compacted rock and well above the present level of the sea. Finally, there were the present-day sediments: sands, silts and muds which were collecting quietly below the water. All of these, *schistus*, hard rock and today's mud and sand, had originally been the same: the rocks formed in the same way as today's sediments. How this had happened Hutton had no idea, but he did realize that it must have taken a great deal of time. The similarities between the present day sediments, the hard Devonian rocks and the recrystallized Silurian *schistus* were as important to him as their present differences. This was an astonishing observation in 1794, and what marked Hutton out as exceptional. Let's return now to Mars, but with Hutton at our shoulder and whispering into our ear.

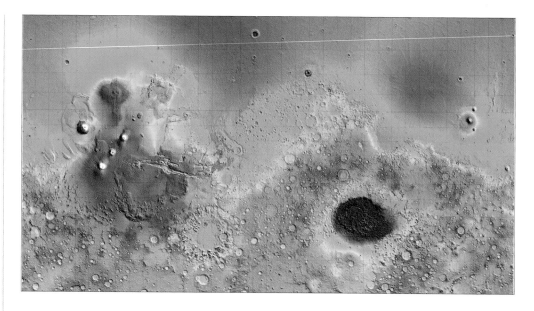

Figure 7.7
The MOLA map (Mars Orbital Laser Altimeter) showing clearly the smooth northern plains and the old, cratered southern Highlands. Also imaged are the straight line of the Valles Marineris, the volcanoes of Olympus Mons and Tharsis Montes, and the craters of Argyre (left) and Hellas. Blue is low, red to white high. (Courtesy of NASA.)

Mars is a most peculiar planet; besides being red, as everyone knows, it is severely lop-sided. The surface of the cratered, Noachian-aged (4.5–3.8Ga) southern Highlands is complex and topographically rough, but on average high: 1–3km above Mars datum. This contrasts with the northern third, which is low, 4.0km below datum, smooth and lightly cratered and so 'young', 2.5–3.5Ga (Hesperian to Amazonian). There is a marked, continuous boundary (the dichotomy) between the two. Planet observers say that the northern lowland is topographically the smoothest area in the whole Solar System, a planetary billiard table. It could be a sedimentary surface (smothering something older), although the American Martian surface landers have found mostly blocks of lava and eddies of red, wind-blown dust in the smooth areas they have examined elsewhere. The 5–7km difference in height between this surface and the Highlands compares with an equivalent on Earth of only about 4.5km between continents and ocean floors, +0.7 to -3.8km. The big difference on Mars is surprising, considering that it is a small planet and only one-quarter the volume of the Earth. Why does this small planet have such noticeably lop-sided scenery? And why is the northern basin so flat? Experts are still searching for satisfactory explanations. As a non-expert, and with a whisper in the ear from Hutton, it is tempting to look at the similarities with Earth, rather than the differences. On its surface, Earth also has divisions; there are continents (30% of the surface) and oceans (70%). They have a difference in age: oceans are young, continents old. They have a difference in topography: the oceans are low and smooth, the continents high and rough. And of course the oceans are full of water, the continents are dry. So why not explain Mars in Earth terms, and make the old, rough, Martian Highlands 'a continent' and the young, smooth Lowlands 'an ocean'? Indeed, some scientists have done just this and even suggested that they can pick out a 'shoreline' surrounding the 'oceanic' lowlands running along the dichotomy. Most Martian geologists (experts on Mars, that is!) have dropped this theory because they cannot explain the hole and cannot find the water. These sound like excuses.

On Earth, crustal differences are the result of continental drift (plate tectonics), and if we let reasoning free for a while, we can suggest some type of 'continent' and 'ocean' formation to explain Martian crustal differences. There is, however, no evidence for classical continental drift on Mars. Actually, 'no evidence' means that there is no magnetic evidence (magnetic stripes) and no tectonic evidence for lateral or other movements, the entire Lowlands being surrounded on all sides by what would be a divergent margin (where new crust is produced): an obvious impossibility. QED, no typical continental drift. This is not being entirely honest. There is vague evidence (possible stripes) for past Martian magnetism (there is effectively none now) and ancient lateral movements may be hard to spot under the smothering of wind dust or other sediment. The differences between the Martian Highlands and Lowlands appear to be exactly the differences between

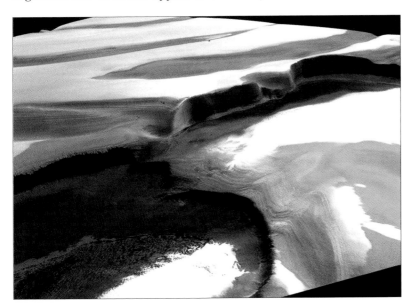

Figure 7.8
This is Mars! Ice and dust at the Martian North Pole. The more images we obtain from Mars the more similarities there are with parts of the Earth (courtesy of ESA).

Earth continents and oceans, which would mean that in the past, Mars had crustal differences and an ocean, the same as on Earth now. The more we look, the more Mars becomes similar to the Earth. Hutton takes another view: Earth has aspects of Mars. When we compare the marks on the bedroom wall, we see that at one point they were at the same level; the Earth and Mars would have both looked the same with continents, large meteorite craters, oceans, an atmosphere and – primitive life?

Mars, of course, is no longer like the Earth and has very ancient surfaces, so any similarities would also be ancient, although the various Martian rovers will tell us more about this. The *schistus* and rocks of Siccar Point are different from the mud and sand on the beach today. Hutton, of course, sees the fundamental similarities. Even so, when we look for similarities to Earth on Mars we are beginning to find them. There are convincing signs of water, and in all probability we will eventually find signs of past, if not present, life. We do not seem to be so successful with Earth's similarities to Mars. Originally our planet would have been as scarred as Mars is now, by very large, very early meteorite craters, and these would have marked our planet permanently. Huge holes, great circular mountainous walls and great outpourings of lava would have accompanied these early impacts. Earth would have been lop-sided while Mars would have been showing signs of life. Perhaps we are missing the clues?

James Hutton presented his theory of the Earth in the face of the many preconceptions of his time; eventually those changed. Two hundred years after him the Apollo missions to the Moon caused another dramatic shift in preconceptions, and whether we recognize it or not, our present global attitude to many things is the result. Finding a link between rocks and reason is, you might think, rather far-fetched, but 200 years ago in Edinburgh, David Hume was the first to notice it. Emotion directs reason, he said. Indeed, there has always been a tie between Man's estate and the way he has explained the natural world about him, as Hume saw. The Greeks imposed their logic on the construction of a nature they understood; for them it was all perfectly ordered. The Incas, on the other hand, sacrificed virgins to a nature they could not understand; it was bewilderingly chaotic, better appeased with what they themselves valued, rather than thought about. In the Industrial Revolution, Hutton considered that the Earth was a mechanical engine, constantly working away. Today, in a global business world, we think of nature in terms of resources, both living and inanimate: the limitations of oil supply; so much drinkable water per person; carbon emissions per car; so many living species. A rather ugly, accountant's view.

So the Earth is not a god's creation, a logical system, a machine, an accountant's balance sheet or even appeased with virgins. Some brave thinkers are beginning (timidly) to think of it as just another planet. But is it? Dr James Hutton, the venerated 'Father of Geology', wrote to his friend George Clerk-Maxwell in 1774 '...the world – the world is [...] no more than a turnip – here's to ye – '. My arse! He was clearly misbehaving.

Further Reading

Books, Pamphlets

Cattermole, Peter. 2001. *Mars, the Mystery Unfolds*. Harpenden, England: Terra Publishing. pp.186. ISBN 1 903544 02 5

Dean, D.R. 1992. *James Hutton and the History of Geology*. Ithaca, NY: Cornell University Press. pp.303. ISBN 0 8014 2666 9

Frankel, Charles. 1996. *Volcanoes of the Solar System*. Cambridge, UK: Cambridge University Press. pp.232. ISBN 0 521 47201 0

Hutton, J. 1795. *Theory of the Earth with proofs and illustrations*. 2 Vols. Edinburgh: W. Creech.

Playfair, J. 1802. *Illustrations of the Huttonian Theory of the Earth*. London: Cadell & Davies.

Trewin, Nigel H. (Ed.) 2002. *Geology of Scotland*. Geological Society London. 4th Edition, pp.576. ISBN 1 86239 126 2

Scientific Papers

Cavosie, A.J., Valley, J.W., Wilde, S.A., E.I.M.F. 2005. Magmatic $\delta^{18}O$ in 4400–3900Ma detrital zircons: A record of the alteration and recycling of crust in the Early Archean. *Earth and Planetary Science Letters* pp.663–681.

Craig, G.Y. and Hull, J.H. 1999. James Hutton – Present and Future. *Geological Society London*, Special Publication 150. pp.184.

Eyles, V.A. and Eyles, J.M. 1950. Some Geological Correspondence of James Hutton. *Annals of Science.* Vol. 7. pp.316–339.

Hutton, J. 1788. Theory of the Earth; or an investigation of the laws observable in the Composition and Dissolution of Land upon the Globe. *Transactions of the Royal Society of Edinburgh.* Vol. 1 (2), pp.209–304. (Read to the Society in 1785).

Jones, J., Torrens, H.S. and Robinson, E. 1994. The correspondence between James Hutton (1726–1797) and James Watt (1736–1819) with two letters from Hutton to George Clerk-Maxwell (1715–1784): Part I. *Annals of Science.* Vol. 51. pp.637–653.

Jones, J., Torrens, H.S. and Robinson, E. 1995. The correspondence between James Hutton (1726–1797) and James Watt (1736–1819) with two letters from Hutton to George Clerk-Maxwell (1715–1784): Part II. *Annals of Science,* Vol. 52. pp.357–382.

Peck, W.H., Valley, J.W., Wilde, S.A. and Graham, C.M. 2001. Oxygen isotope ratios and rare earth elements in 3.3 to 4.4Ga zircons: ion microprobe evidence for high $\delta^{18}O$ continental crust and oceans in the Early Archaean, *Geochimica et Cosmochimica Acta,* 65, No. 22, pp.4215–4229.

Wilde, S.A., Valley, J.W., Peck, W.H. and Graham, C.M. 2001. Evidence from detrital zircons for the existence of continental crust and oceans on the Earth 4.4Gyr ago. *Nature,* 409, pp.175–178.

Chapter 8   THE FUTURE

The Future

So what is the future of the Highlands, and indeed the planet as an entity? We have in this book spoken about some aspects of its past and what it has to teach us. These have been selected by the authors' love for just a small part of the crust, but certainly a very special part. Today we have more power than at any time in the long history of the Earth to knowingly influence our collective future. As a global community we struggle to agree how the sum total of all of our activities feeds back to the evolution of our planet and peoples.

We must learn to live, understand and work with our planet, not just treat it as an unfeeling mass. That is not just a few 'experts' understanding and having empathy, but most of us. Only in that way will its needs be taken into account in our decision-making processes. If you have yet to read *The Living Mountain* by Nan Shepherd, then try to do so (Canongate Books, 2011, ISBN 978 0 85786 183 2). The Cairngorms were very much living mountains to her, and as she said:

The thing to know grows with the knowing.

There are still many things for us to find out about our planet and how we can continue our symbiotic relationship. We all depend on this planet, and if it ceases to function fully because of our interventions, then much of the biodiversity is in for a hard time. That doesn't mean that we cannot continue to use the resources it has, but we have to know the implications of that usage for the Earth and ourselves.

The Highlands are stunning in scenery, special in geology, unique in history and hugely rich in archaeology. They are the last refuge for many wild birds, animals, plants and trees: a very rare place of wildness. And they are almost empty of people – which is, of course, why they are a natural refuge. Follow the west wind from the Outer Isles across the Minch and then to the mountain fastness of Sutherland – like a soaring Golden Eagle: the land is yours. For man, beast and bird it is a freedom that is the breath of life, the freedom of inspiration.

Living and working in the Highland landscape has always been hard; it is always a matter of balancing priorities. The natural resources must be worked in a sustainable way, and preferably controlled by those who live and work here. There is a long history of people and organizations coming from outside the Highlands with the best of intentions, but not fully understanding the unique nature of the area and how to get the best outcomes for everyone. These often end in longer-term failure. The proposal for a UK spaceport at A'Mhoine, Sutherland, just outside the geopark boundary, has the potential for positive outcomes for the communities on the north coast. It may be the only place in Europe where satellites could be launched by vertical rockets, and would provide much-needed work, including some highly skilled jobs. The sorts of numbers being suggested are 130 at Forres and 20 locally in Sutherland. It does, however, require the sacrifice of some of the natural environment, and hence the need to balance priorities. It is a development in the early stages of planning that must now involve the local population in setting some of those priorities and opportunities.

Even in this apparently wild and remote area, the landscape as we see it already owes some of its shape to modification by man. Yes, the skeleton is defined by the rock types and structures, but laid on that is a blanket of moraine, soil and vegetation. The soil is mostly thin, and apart from where limestone or Fucoid Beds outcrop it is mostly quite acidic. The ground is often waterlogged because the rocks are crystalline and prevent water drainage, and the peat that has built up retains the water. You may think that the vegetation you see has always been the same, but natural climate changes and the hand of the human population have modified it. The climate was drier over 5000 years ago and the population very low, and hence woodland was much more extensive. It never extended to the mountain peaks but did occupy the lower-lying slopes as well as the valley bottoms. In the peat bogs you will see the tree stumps preserved of some of this woodland, which grew when there was less rain and fewer people to cut the trees down. Many stumps are around 5000 years old and have been unable to decay because of the saturated conditions.

In recent years the remoter parts of the Highlands have seen a reduction in the younger-aged population and hence of the vitality of the community. In the Highlands a community must utilize all the resources it has, especially the human resources, if it is to survive in a meaningful way. Only time will now tell whether our remote Highland communities can continue to flourish on this unique landscape, and what compromises will now be needed for this to happen. The world as a whole must work on the wider dangers for our planet and its ecosystems.

Index

absolute age dating 5, 80
Acasta Gneiss 178, 206, 209
Achanarras Fish Bed 75–76, 80–89
Achnasheen, Younger Dryas flood deposits
      154–155
acid rain 31
acritarchs 19, 22, 24, 25
Agassiz, Louis (1807-1873), ice age 140
albedo, oceanic ice 31
algae
   blue-green 19
   Orcadian Lake 82, 84
   see also acritarchs; cyanobacteria
Allt Poll an Droighinn river 53, 54
A'Mhoine, UK spaceport 219
anaerobes 20
animals 165–166
annelids 26
anoxia, Orcadian Lake 82, 84
Archaean 3, 5, *178*, 179, 190
Ardnamurchan 107
   volcano 116
Ardvreck Castle 52
argon isotopes 207
Arran 107, *111*
arthropods 26
Askival 107, 108, *117*, *119*
Assynt 40, 46–47, 50
   Bone Caves 166
   glaciation 156
Assynt Crofters' Trust logo *32–33*
Assynt Window *48*, 50, 53, 54, 55, 57, *58*,
   68–69
atmosphere 24
   age of 207

bacteria 17, 19, 20, 21
   cyanobacteria 19–21, 24
banded ironstone formations 20, 24
Barringer Meteorite Crater 177
Basal Quartzite 53, 56
basalt
   lunar 204
   Tertiary 107, 110–112
   see also flood basalt
Bay of Stoer 3, 7, *9*, 12
beetles, Younger Dryas Stadial 148–149
Beinn Dearg Mor 151–152
Beinn nan Stac 117–118
Ben Arnaboll 63
Ben More Assynt 50, 53, 57
   glaciation 156, 157
Ben More Mull 107
Ben More Thrust 58

biostratigraphy 78–79
Bone Caves 166

brachiopods 26
British Tertiary Volcanic Province 107, *130*
Buckland, William (1784-1856) 140
Burgess Shale 25–26

Caithness Flags 80–81
calcium carbonate 21, 28, 82
   skeletons 28
Calda House 52–53
Calloway, Charles (1838-1915), Highlands
      Controversy 67–68
Cambrian explosion 25–28
   and Snowball Earth 29–34
Cambrian Quartzite 51, 57, 156
Cambrian-Ordovician succession *53*
Canisp 2
   ice scratches 158
Canna, lava 119
canyons, Lewisian-Torridonian
      unconformity 8, 9–11, *10*
cap carbonates 30, 31
Cape Wrath 174–175
   Lewisian Gneiss 178, 179
carbon isotopes 22, 30, 31
carbonate, Achanarras Fish Bed 82
Charnia masoni 27
chordates 26
Chrons *133*, 134
Clachtoll
   canyon 8, *11*
   Torridonian sediments 7, 8
Clar Loch Mor 187, *188*
Clashnessie canyon *11*
climate change 141, 150, 160–169, 219
Cnidarians 27
Coelacanth 90–91, 92, 97, 99, 100
Coire Dubh 108, 109, 110, 117
Conival 50, 53, 54, 57
   glaciation 156, 157, 158
   ice scratches 158
continental crust 190
continental drift 183, *184*
convection currents 182–183
convergent margin 184
   Archaean 190
core 181, 206
Creag a' Mhàll 191
Crossopterygii 86, 89
crust 181, 183, 190, 206
Cul Beag 2, *3*
Cul Mor 2, *3*, 56

Culicoides impunctatus 149, 167, 186
cyanobacteria 19–21
  oxygen production 19, 20, 24

degassing, mantle 207
Desmarest, Nicholas (1725-1815) 112
desiccation cracks 13–14, 81
Devensian glaciation 146, 156–157
Devonian
  Old Red Sandstone 80
    fish beds 75–76, 80–89
Diabaig, fossil rain drops 14–15
Dickinsonia 27
digits 92, 94, 95
dip 7, 8, 9
Dipnoi 86
Dipterus valenciennesi 85–87
Dismal Lakes, stromatolites 21
divergent margins 184
DNA 99
dolerite dykes *188*, 190–194
Drosophila melanogaster 94, 98–99
Dryas octopetala 148
Durness Limestone 49, 53–54, 55, 59
  Silurian fossils 59–61
dust storms 14
dykes 110, 131
  dolerite *188*, 190–194

Earth
  age of 177, 178
  atmosphere 24, 207
  comparison with other planets 203,
    206–212
  cooling 180, 185, 195–196
  formation of 179–180
  from space *202*
  future of 218–219
  internal structure 180–185, 206
  magnetic field 133–134, 206
  meteorite craters 177
  orbital changes 162
  origin of life 22–23
  Precambrian 178
  environment 22
earthquakes
  at volcanoes 115, 116
  seismic waves 180, *181*
eccentricity 162
echinoids 26
Ediacaran fauna 26–27
Eigg, volcano 118
Eilean Dubh Limestone 53, 54
embryology, limb development 93–95
Erlend volcano 129–130
erosion, ice 141, 147, 151, 156, 157–159
eukaryotes 19, *20*, 22, 24, 25

europium, lunar 204–205
evolution 89, 90, 91, 92, 94–95, 98–100
  Cambrian explosion 25–26

Faroe Islands 129
Faujas de Saint-Fond, Barthélemy (1741-
  1819) 112
faulting, work of Nichol 62–63, 66
Fedonkin, Mikhail (b.1946), *Dickinsonia* 27
fish
  fossilized *see* Achanarras Fish Bed
  nitrogen problem 96–97
  salinity problem 95–96
fjords 155
flash floods
  Stoer 10, 12, 14
  Younger Dryas Stadial 151–156
flood basalt 119, 126–129, 130–131
flooding
  Stoer 10, 12, 14
  Younger Dryas 151, 152–156
foraminifera 145
forests 160–162, 164, 165
Fortingall Yew 161
fossilization 84–85, 86–87
fossils
  in dating sediments 4, 78–79
  in Universal Sequence 77
Fucoid Beds 52, 53, 56

gabbro, layered, Rum 110, 117, 118,
  119–123
Garvie Island 175
Geikie, Archibald (1835-1924)
  collaboration with Murchison 66–67
  glaciation 141
  Highlands Controversy 44, 68–69
  Life of Sir Roderick I. Murchison (1875)
    15, 62, 63, 67
Geikie, James (1839-1915), glaciation 141
gene switches 94
genes 94, 98–100
Geological Survey, Highlands Controversy
  68–69
geological time 4–5, 179
Ghrudaidh Limestone 52, 53, 54, 56, 57
Giant's Causeway 111, 112
glaciation 140–141, 147–169
Gleann Dubh 53, 54
  faults 58
  glaciation 158
Glencoul Thrust 54, 58
Global Positioning System (GPS) 113, 114,
  183
global warming 141, 142, 150
Glytolepis paucidens 88
gneiss 42, 43, 176, *see also* Acasta Gneiss;

Isua Gneiss; Lewisian gneiss
Gorm Loch 186
  geological map 187, *188*, 189, 190
Grand Canyon 9, 10–11
granulite facies 189
gravity anomalies 16–17
Great Wood of Caledon 160–161, 162, 165
greenhouse gas 31
Greenland Ice Core Project (GRIP) 144, 146
Greenland Ice Sheet Project (GISP) 144, 146
Gruinard Bay
  dykes 193
  sea-level change 159
Guettard, Jean-Etienne (1715-1786) 112

Haeckel, Ernst (1834-1919), embryology 93–94
Hallival 106, 108
Hallucigenia sparsa 26
Hawaii, vulcanology 113–116, 123
Hicks, Henry (1837-1899), Highlands Controversy 67
Highlands
  climate 140–169, 219
  glaciation 146–169
  ice *140*, 141, 143
  sustainable future 218–219
Highlands Controversy 40–45, 58–69
Himalayas, thrust faults 49
Horne, John (1848-1928) 45–46, 47, 55, 69
  Lewisian Gneiss 176, 185–186
hot spots 130–131, 133
Hox genes 94, 99
human activity, impact on flora and fauna 164–166, 219
Hume, David (1711-1776) 202, 212
Hutton, James (1726-1797) 200–202, *201*, 212, 213
  Siccar Point 209
  Theory of the Earth (1795) 200
  Vulcanism 111, 179
hydro-electric power 142

ice ages 140, 141, 143
  future prediction 162, 163
  Snowball Earth 29
ice cores 143–144, 146
ice scratches 157–159
ice-shattering 157
Iceland, mid-Atlantic Ridge 133, 134, 184
ichthyolite 83
Inchnadamph Hotel 45, *46*, 47, 50
iridium 16
Irish Elk 166, *168*
iron oxides 20, 24

banded ironstone formations 20, 24
isoclinal folds 188–189
isostatic rebound 159
Isua Gneiss 178, 209

Jameson, Robert (1774-1854), Neptunism 112

karst 53–54
Kilauea 113–116, 121–122
Kinloch Castle 108
Knockan Crag 50, 53, 57
  National Nature Reserve 55–56
Kyle of Durness 174–175

lagerstätten 26
laminites, Achanarras Fish Bed 81–83
Lapworth, Charles (1842-1915), solution to Highlands Controversy 68–69
Latimeria chalumnae 90–91
Laurentia 195
lava 107, 111–112, 121, 129, 130–131
  Kilauea 114–115
  Rum 118–119
  Skye 119, 126–127
Lewisian Gneiss 3, 6–8, 47, 50, 65, 175–176, 179, 185–196
Lewisian-Torridonian unconformity 3, 6–11, *8*, *10*, 50, 53
life, origin of 22–23
limbs, evolution 92–95
limestone *42*, 43, 47, 78
  see also Durness Limestone; Eilean Dubh Limestone; Mountain Limestone
Little Ice Age 167, 168, 169
Loch Assynt *40–41*, 45, 50
Loch Lomond Readvance *see* Younger Dryas Stadial
Loch Maree, glaciation *148*, 150
Loch na Sealga 152–153
Loch Scresort 108, 109
lungfish 86
Lyell, Charles (1797-1875)
  Principles of Geology (1830) 112
  'the present is the key to the past' 11, 160, 180

Macculloch, Dr John (1773-1835) 3
  geological account of Scotland 42
  geological section near Stoer 3
  geological section Sutherland *42–43*
  Highlands Controversy 41, 43–44, 57, 59
machair 7, 12
MacLeod's Tables 127
magma
  Kilauea 114, 115
  Rum 125

magma chamber 116
  Kilauea 114, 121, 122
  Rum *109*, 110, 116, 117–118, 120
magnetic field 133–134
mantle 181–182, 206
  convection currents 182–183
mantle plumes 130–131
Manx Shearwater 106–107
Mars
  comparison with Earth 211–212
  meteorite craters 208–209
  ripples 23
  search for life 23
  surface *12–13*, 208–209, 210–211
Medieval Optimum 166, 168, 169
megagrooves 158–159
meiosis 25
metamorphism
  Lewisian Gneiss 6, 175–176, 189, 195
  Scourian 189–190, 191, 194
metazoans 25–26
  coelomic 27
meteorites
  carbonaceous chondrites 177
  stony 124–125
meteorite impact
  Torridonian 15–17
    ejecta 15–17, 131–132, 178
microbial mats 17, 20
  see also stromatolites
mid-Atlantic Ridge 133, 184
midges 149, 167, 186
Milankovitch cycles 162, 163
Miller, Hugh (1802-1856) *74*, 75
  The Old Red Sandstone (1841) 74, 83, 85,
    87, 88, 89, 91, 98, 99, 100
mitosis 19, 25
Moine Schist *45*, 55, 57, 60, 64, 65
Moine Thrust 40, *45*, 47–49, 55, 57–58
  Cambrian-Ordovician succession *53*
molluscs 26, 78
Moon
  lava *maria* 208
  meteorite craters 177, 208
  rock samples 203–205
    europium 204–205
    REEs 204
  surface age 5, *6*, 177, 204
moraine, terminal 147, *148*, 152
Mount Pinatubo *132*
Mountain Limestone 59, 78
Muck, volcano 118
mud cracks, fossilized 13–14, 81
Mull 107
Murchison, Sir Roderick Impey (1792-
  1871) *44*
  Durness Limestone Silurian fossils 59–61

Highlands Controversy 41–42, 44, 57,
  58, 59–67
mylonite 49, 55, 65

nappes 57
Neoproterozoic 26
  glacial tillites 29, 30, 33
Neptunism 42, 111, 112
New Red Sandstone 78
Nichol, James (1810-1879)
  Durness Limestone Silurian fossils 59–61
  faulting 62–63, 66
  opposition to Murchison 44, 57, 62–67
North Atlantic Ocean 107, 130–134
North Coast 500 55
North West Highlands UNESCO Global
  Geopark 55–56
nunataks 157

Ocean Drilling Program (ODP) 146
oceanic crust 190
oceanic ice 31
Old Man of Storr 127
Old Red Sandstone 43, 60, 61
  work of Miller 74, *78*
ontogeny 93
Orcadian Lake 81–82
  mass death 83–84
oxides 20
oxygen
  cyanobacteria 19, 20, 24
  earth's atmosphere 24
oxygen isotopes, as proxy for past
  temperature 144–147

P waves *181*
palaeomagnetism 29, 133–134
Palaeospondylus gunni 87–88
Peach, Benjamin Neeve (1842-1926)
  45–46, 47, 55, 69
  Lewisian Gneiss 176, 185–186
Peach, Charles (1800-1886), Durness
  Limestone Silurian fossils 59–61
phosphate 28
photoautotrophs 19
photosynthesis 19, 20, 21
phylogeny 93
Pilbara Supergroup, stromatolites 19, 22
Pinus sylvestris 161
Pipe Rock 43, 51–52, 53, 54, 56, 57, 65
  'worm burrows' 51–52, 63
plate tectonics 183–185
polarity changes 133–134
Pre-Cambrian 3–4
  see also Precambrian
Precambrian 5–34
  evolution 21–22

timescale 4–5
precession 162
Primary rock 42, 43, 58, 77
prokaryotes 19, *20*
Proterozoic 5–6, 17
Proterozoic-Cambrian boundary 26, 33, 51, 53
Pterichthyodes milleri 88

quartz, shocked 16
Quartz Rock 43, 57, 60–61, 63, 64, 65
  see also Cambrian Quartzite
quartzite, Assynt 43, 54
Quinag 2, *3*, 51
Quiraing 126–127

rain drops, fossilized 14–15
raised beaches 155–156
Ramsay, Andrew (1814-1891) 64
rare earth elements (REEs) 123–125
  heavy (HREEs) 124, 125
  light (LREEs) 124, 125
  Moon rock 204
reefs, bacterial 21
Reidite 16, 132
ring faults 108, *109*, 110, 117
ripples, fossilized 4, 13, 81
Rockall 128–129
Rodinia supercontinent 30
Romer, Alfred Sherwood (1894-1973), cranial evolution 95, *96*
Rum 107, 108
  volcano 108–110, 116–119, 122–123, 125–126
    erosion 119
    eruption patterns 122
    see also gabbro, layered; magma chamber, Rum
Rum Cuillin 106

S waves *181*
St Kilda 107
salinity, problem for fish 95–96, *97*
salmon, salinity problem 95–96, *97*
Salterella 52
Salterella Grit 52, 53, 56
Sarcopterygii 86
schist 42, 43, 57, 60, 189
  see also Moine Schist
Scourie, Lewisian Gneiss 186, 188
Scourie dolerite dykes *188*, 190–194
sea lochs 155
sea-level change 146, 155–156, 159
Secondary rock 42, 43, 58, 77
sedimentology 4
seismic waves 180, *181*
Shark Bay, stromatolites 19, 21

Siccar Point, Devonian-Silurian unconformity 209
Silurian 42
  Durness Limestone fossils 59–61
  work of Murchison 61–64
skeletons, calcium carbonate 28
Skiag Bridge 51, *52*
skull, evolution 95, *96*
Skye
  Cuillins 107
  Quiraing lava 119, 126–127
slime *see* microbial mats; stromatolites
Smith, William (1769-1839), biostratigraphy 78
Snowball Earth hypothesis 29
  and Cambrian Explosion 29–34
  melting 31
  survival of life 32
  weather 31
'soda ocean' 21
Sole Thrust 48, 52, 58
specialization, in geoscience 142–143
Split Rock logo *32–33*
Sprigg, Reginald C. (1919-1994), Ediacaran fauna 27
stade 147
Staffa 107, 111, 112
Stoer 2, 6
  acritarchs 25
  canyon 9–10, 11
  Lewisian-Torridonian unconformity 6–8, 9
  mud cracks 14
  stromatolites 21–22
Strath na Sealga 151–152
stratigraphic column 4, *5, 79*
stratigraphy 4, 77
strike 7, 8
stromatolites 18–22, 24, 28
  Stoer 21–22
Stronchrubie cliff 50, 53, 54
  glaciation 158
subduction 184
  Lewisian Gneiss 190
Suilven 2, *3*
  glaciation 158
superposition, law of 44
Sutherland
  Moine Thrust 40, *45*, 47–49
  work of John Macculloch *42–43*
Swaziland Supergroup, stromatolites 19, 22
synform 189

Tertiary
  lava field 129, 130–131
  volcanoes *106–7*, 107

Tertiary rock 42, 77
Thingvellir 133, 134
thrust faults 45, 48–49, 53–54, 57–58
tillite 29, 30, 33
  palaeomagnetism 29
tilt 162
tiltmeters 113, 114, 115
Torridonian 3–34
  meteorite impact 15–17
  sedimentary rocks 6–7, 12–14, 43, 47,
    51, 60–61, 65
    Rum 109
      magma chamber 117, 118
tourism 55–56
Traligill river 53
transform faults 184
traps *126*, 127
trilobites 52
tuff 80, 132–133

ultrabasic rock 189
unconformity
  Devonian-Silurian 209
  Lewisian-Torridonian 3, 6–11, *8, 10*
Universal Ocean 42, 111
Universal Sequence 42–43, 77
uraninite 24
urea 97

Varangian ice age 29
varves 82
vegetation, indicator of climate change
  160

volcanoes
  ash age dating 80
  eruption 122
  greenhouse gases 31
  historical theories 110–112
  modern studies 113
  monitoring 113–114, 121, 123
  preservation of life during Snowball
    Earth 32
  Tertiary *106–7*, 107
    Rum 108–110, 116–119, 122–123,
      125–126
Vostok Ice Core 146
Vulcanism 111, 112

Walcott, Charles Doolittle (1850-1927),
  'Cryptozoon' 18–19
Wegener, Alfred (1880-1930), continental
  drift 183, *184*
Werner, Abraham Gottlob (1749-1817)
  origin of rocks 111
  Universal Ocean 111
  Universal Sequence 42–43, 77–78
wind erosion 14
wind farms 142, 167
'worm burrows' 15
  Pipe Rock 51–52, 63

Younger Dryas Stadial *146*, 147–156

zircon, Jack Hills 178, 206